RESEARCH REPORT

A Hotter and Drier Future Ahead

AN ASSESSMENT OF CLIMATE CHANGE IN U.S. CENTRAL COMMAND

Michelle E. Miro, Flannery Dolan, Karen M. Sudkamp, Jeffrey Martini,
Karishma V. Patel, Carlos Calvo Hernandez

NATIONAL DEFENSE RESEARCH INSTITUTE

For more information on this publication, visit **www.rand.org/t/RRA2338-1**.

About RAND

The RAND Corporation is a research organization that develops solutions to public policy challenges to help make communities throughout the world safer and more secure, healthier and more prosperous. RAND is nonprofit, nonpartisan, and committed to the public interest. To learn more about RAND, visit www.rand.org.

Research Integrity

Our mission to help improve policy and decisionmaking through research and analysis is enabled through our core values of quality and objectivity and our unwavering commitment to the highest level of integrity and ethical behavior. To help ensure our research and analysis are rigorous, objective, and nonpartisan, we subject our research publications to a robust and exacting quality-assurance process; avoid both the appearance and reality of financial and other conflicts of interest through staff training, project screening, and a policy of mandatory disclosure; and pursue transparency in our research engagements through our commitment to the open publication of our research findings and recommendations, disclosure of the source of funding of published research, and policies to ensure intellectual independence. For more information, visit www.rand.org/about/principles.

RAND's publications do not necessarily reflect the opinions of its research clients and sponsors.

Published by the RAND Corporation, Santa Monica, Calif.
© 2023 RAND Corporation
RAND® is a registered trademark.

Library of Congress Cataloging-in-Publication Data is available for this publication.
ISBN: 978-1-9774-1237-9

▌ABOUT THIS REPORT

THIS REPORT PRESENTS analysis and characterizes how climate change will affect the physical environment in the U.S. Central Command (CENTCOM) area of responsibility (AOR) in 2035, 2050, and 2070, highlighting the climate hazards that could most significantly affect regional populations and influence the potential for conflict. To accomplish this, we used climate data analysis in addition to a qualitative literature review.

This report is the first in a series stemming from a larger project considering the impacts of climate change on the security environment in the region. The second report, *Pathways from Climate Change to Conflict in U.S. Central Command*, presents an analysis of the causal pathways from climate change to conflict, including cases in which those pathways have played out in the CENTCOM AOR. *Conflict Projections in U.S. Central Command: Incorporating Climate Change*, the third report, presents a range of forecasts of future conflict in the region, with climate change incorporated as one driver of that conflict. The fourth report, *Mischief, Malevolence, or Indifference? How Competitors and Adversaries Could Exploit Climate-Related Conflict in the U.S. Central Command Area of Responsibility*, presents an analysis of how competitors—China, Russia, and Iran—might attempt to exploit climate-related conflict in the region. And the final report, *Defense Planning Implications of Climate Change for U.S. Central Command*, presents an analysis of "off-ramps" to avoid climate-influenced conflict and the operations, activities, and investments that are needed for CENTCOM to be prepared to mitigate given climate impacts on the security environment. The primary audience for these reports is CENTCOM planners, intelligence officers, and leadership. The research reported here was completed in May 2023 and underwent security review with the sponsor and the Defense Office of Prepublication and Security Review before public release.

RAND National Security Research Division

This research was sponsored by CENTCOM and conducted within the International Security and Defense Policy Program of the RAND National Security Research Division (NSRD), which operates the National Defense Research Institute (NDRI), a federally funded research and development center sponsored by the Office of the Secretary of Defense, the Joint Staff, the Unified Combatant Commands, the Navy, the Marine Corps, the defense agencies, and the defense intelligence enterprise.

For more information on the RAND International Security and Defense Policy Program, see www.rand.org/nsrd/isdp or contact the director (contact information is provided on the webpage).

Acknowledgments

We thank Scott Stephenson of the RAND Corporation and James Yoon of Pacific Northwest National Laboratory for their thorough review of this report. Their review and constructive commentary was essential to the publication of this report.

We thank Chuck Story for his support generating the maps contained in this report. We also extend deep gratitude and appreciation to Jessica Arana, Kristen Meadows, and Chandra Garber. Their respective design and research communications expertise ensured that this report is accessible to readers at any level of climate literacy. We also thank Valerie Bilgri for her detailed review and editing of this report. Finally, we thank Katherine Anania for her assistance in the early stages of the literature review, which was critical to identify the climate hazards on which this analysis rests.

▌CONTENTS

FIGURES AND TABLES

Figures

Tables

REPORT 1
Climate Hazards and Impacts
- Identify climate hazards
- Conduct climate analysis

REPORT 2
Conflict Pathways

REPORT 3
Conflict Projections

REPORT 4
Adversarial Responses

REPORT 5
Defense Planning Implications

‖KEY FINDINGS

THE BROADER AIM of this project centers on characterizing the relationships between climate change and conflict to inform operational and longer-term decisionmaking by the U.S. Central Command (CENTCOM). To set the stage for this analysis, we conducted a regional climate assessment that quantifies changes in key climate hazards—extreme heat, extreme cold, drought and long-term dryness, extreme precipitation, dust storms, and coastal inundation—and the effects of these hazards on food and water security.

- Nearly all aspects of the CENTCOM area of responsibility (AOR) face the compounding effects of high temperatures, drought, and long-term dryness. These effects are accelerating across the CENTCOM AOR, which spans from Egypt through the Levant and the Arabian Peninsula and from Iran to Central Asia and Pakistan.
- As the AOR becomes drier, existing water resources will become scarcer. This could be particularly acute across the AOR compared with other parts of the world, given existing water scarcity issues and the high degree of tension that already exists around shared water resources.

- More frequent and more severe extreme heat events coupled with drier conditions will make agricultural production more difficult throughout much of the AOR. This will likely be the case even in regions where warmer temperatures are lengthening the growing season, such as Central Asia.
- Many countries in the AOR, such as Yemen, Oman, and Pakistan, are experiencing aridification in concert with increases in extreme precipitation, heightening the risk of flash flooding.
- There are a few key hot spots that will see additional compounding hazards. Southern Iraq—including Basra, Maysan, and Dhi Qar governorates—is vulnerable to sea level rise, surface water losses, and extreme heat. Additionally, Alexandria and Port Said in Egypt face risks from sea level rise and declines in surface water availability from the Nile.

Finally, the findings in this report are drawn from four climate models and two emissions scenarios to represent future climate uncertainty. The magnitude of uncertainty can have enormous implications for planning, making it critical to consider the implications of this uncertainty in decisionmaking processes.

CHAPTER 1

INTRODUCTION

CLIMATE CHANGE IS increasingly becoming a major disruptor of human and natural systems. In some areas, summer temperatures are quickly rising, droughts are deepening, and heat waves are lengthening and getting hotter. Such changes will place pressure on scarce water resources, threaten food security, disrupt fisheries, and result in direct health consequences, among other impacts. These effects can produce secondary and tertiary impacts on human systems that may destabilize societies, economies, or governments. However, these dynamics are highly complex and deeply uncertain, and the pathways from climate changes to societal disruptions that lead to conflict remain poorly understood and an area for continuing research. Still, decisionmakers must plan and act in

the near term to reduce future climate-induced risks to physical and human systems.

This report centers on characterizing the relationships between climate change and conflict to inform operational and longer-term decisionmaking by U.S. Central Command (CENTCOM). To that aim, Figure 1.1 provides an example of a climate-change-to-conflict pathway, illustrating potential direct and indirect effects on environmental and human systems that can lead to societal instability. In this report, we focus on understanding the first two steps: characterizing climate change and its impacts on environmental systems, with an emphasis on impacts that are relevant to food and water security. Subsequent reports will address the third and fourth

Figure 1.1. Example Impacts of Climate Change

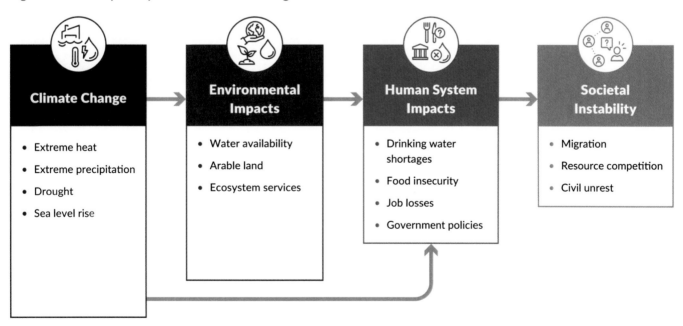

SOURCE: Adapted from Jürgen Scheffran, P. Michael Link, and Janpeter Schilling, "Theories and Models of the Climate Security Link," Working Paper CLISEC-3, University of Hamburg, Research Group Climate Change and Security, January 2009.

steps of the pathway, along with opportunities for intervention to mitigate the risk of conflict occurring.

We adapted our understanding of the cycle of climate change hazards through impacts to the physical and security environments, as shown in Figure 1.1, to illustrate our report series. Through five reports, we identify the climate hazards that will have the most impact on the Middle East and Central and South Asia, illustrate the climate-hazard-to-conflict pathway, and highlight how CENTCOM will have to be prepared to mitigate the risk of climate-related conflict and support the resiliency of partner nations in the region. Figure 1.2 illustrates the report series.

This study used the CENTCOM area of responsibility (AOR), shown in Figure 1.3, as the geographic boundary for our research. The AOR consists of 21 countries stretching from Egypt in the west through the Levant, the Arabian Peninsula, and Iran to Central Asia and Pakistan in the east, spanning two continents and multiple subregions. Although the climate is primarily hot and dry, topographic variability and local climate conditions mean that there is a high degree of geographic variability in what natural hazards affect each country and how those countries are shifting with climate change. However, no comprehensive climate assessment of the AOR exists, even though such information is critical to understanding how climate change could trigger societal disruptions and, ultimately, conflict.[1]

We thus focused our analysis on conducting a regional climate assessment that quantifies changes in key climate hazards and the effects of those changes on food and water security. We started with a literature review to identify a set of climate hazards of importance to the region. We then obtained and analyzed a climate model output to characterize how climate hazards are projected to change across the CENTCOM AOR in 2035, 2050, and 2070. We also used a global land surface and a global hydrologic model to study how changes in climate hazards would affect crop production and water availability. A selection of our most significant findings is described in this report, organized by the three CENTCOM subregions: the Levant and Egypt, the Central Gulf, and Central and South Asia.

Figure 1.2. Progression of Reports in this Series

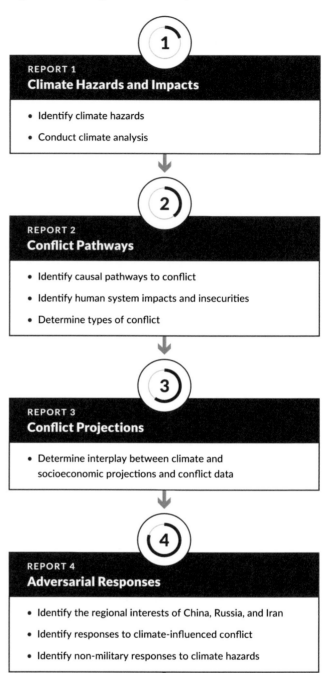

1

REPORT 1
Climate Hazards and Impacts

- Identify climate hazards
- Conduct climate analysis

2

REPORT 2
Conflict Pathways

- Identify causal pathways to conflict
- Identify human system impacts and insecurities
- Determine types of conflict

3

REPORT 3
Conflict Projections

- Determine interplay between climate and socioeconomic projections and conflict data

4

REPORT 4
Adversarial Responses

- Identify the regional interests of China, Russia, and Iran
- Identify responses to climate-influenced conflict
- Identify non-military responses to climate hazards

5

REPORT 5
Defense Planning Implications

- Identify off-ramps to conflict and requirements
- Identify likely intervention types
- Analyze likely costs associated with interventions

Figure 1.3. The U.S. Central Command Area of Responsibility

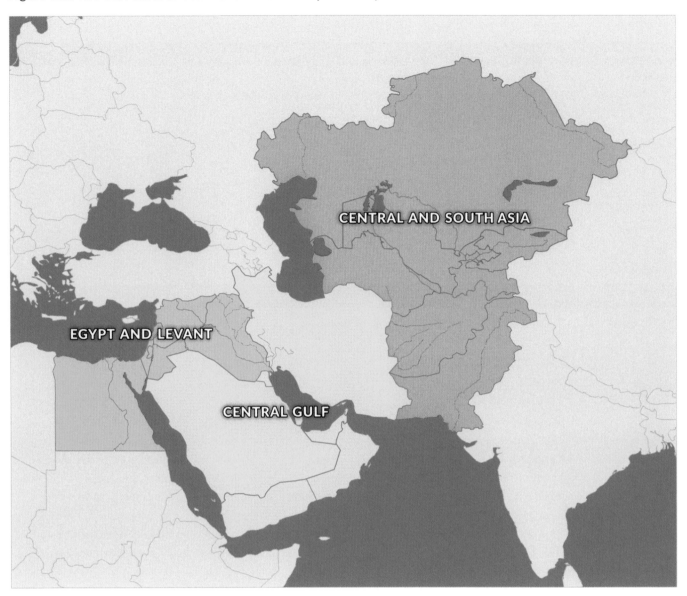

CENTRAL AND SOUTH ASIA

EGYPT AND LEVANT

CENTRAL GULF

SOURCE: RAND-designed graphic based on email communications with CENTCOM, October 12, 2022.

Methods

Our analysis of climate change in the CENTCOM AOR involved three main efforts: literature review, data acquisition and processing, and data synthesis. To identify the primary climate hazards relevant to each subregion, we drew from a literature review of major climate change studies in the region (see Appendix A for more details). Table 1.1 presents the climate hazards and environmental impacts we examine in this report and the metrics used to quantify those changes over time. Here, we use the term *climate hazards* to represent the first-order impacts associated with climate change (e.g.,

lower precipitation). *Environmental impacts* are defined as the second-order effects that result from climate hazards (e.g., decreased surface water availability).

To calculate the above metrics, we used data from an archive of gridded general circulation model (GCM) outputs (see Appendix A for more details on data sources).[2] Because the impacts of climate change are highly uncertain, especially decades into the future, we analyzed the output from four GCMs that were based on two Representative Concentration Pathways (RCPs), which represent plausible emissions

Table 1.1. Climate Hazards and Environmental Impacts Examined for the U.S. Central Command Area of Responsibility

Climate Hazards	Metrics Assessed
Extreme heat	• Annual number of days with temperatures above 95°F • Annual number of days with heat and humidity that pose a risk to human health
Extreme cold	• Annual number of days with temperatures at or below 32°F
Drought and dryness	• Maximum consecutive dry days annually • Annual maximum consecutive dry days • Aridity (degree of dryness)
Extreme precipitation	• Annual maximum precipitation in 24 hours
Dust storms	• Annual number of windy days • Aridity (degree of dryness)
Coastal inundation	• Sea level rise

Environmental Impacts	Metrics Assessed
Water resources	• Groundwater pumping and recharge • Surface water availability
Agriculture	• Crop yield
Fisheries	• Fish stock

trajectories developed by the climate modeling community for uncertainty analysis.[3] For our analysis, we used the high emissions scenario (RCP 8.5) and a moderate warming scenario (RCP 6.0).[4] These scenarios are meant to show what climate changes may occur in the absence of stringent global climate policy and to present a more pessimistic future for planners to consider (RCP 8.5).

Because we calculated metrics from the output of four GCMs and two emissions scenarios, each grid cell within the AOR contains eight future estimates for a given metric. To characterize the variety of plausible climate changes and their impacts, we report the median value of these eight estimates and note the maximum and minimum values where relevant. In addition, all values shown in the report represent the historical baseline or projections in 2035, 2050, or 2070. Finally, except for mapped results, all findings reflect country-level averages.

Endnotes

[1] Although international climate assessments, such as the Intergovernmental Panel on Climate Change's (IPCC's) Sixth Assessment Report, are available, these are not at a resolution that is sufficient to understand country- and subcountry-level climate changes or identify areas of concern for the region.

[2] The grid cells for all models were 0.5 by 0.5 degrees, which is approximately equivalent to 55 km by 55 km at the equator.

[3] GCMs are complex climate models that solve systems of differential equations based on the basic laws of physics, fluid mechanics, and chemistry. This uncertainty stems from the unknown trajectory of future emissions and from uncertainty in our understanding of the climate system.

[4] To represent the high emissions scenario, we used RCP 8.5. RCP 8.5 is characterized by an average global warming of approximately 3.7°C (6.7°F) by the end of the century, a low global effort to curb emissions,

and primarily nonrenewable sources of energy generation at the global scale (Detlef P. van Vuuren, Jae Edmonds, Mikiko Kainuma, Keywan Riahi, Allison Thomson, Kathy Hibbard, George C. Hurtt, Tom Kram, Volker Krey, Jean-Francois Lamarque, Toshihiko Masui, Malte Meinshausen, Nebojsa Nakicenovic, Steven J. Smith, and Steven K. Rose, "The Representative Concentration Pathways: An Overview," *Climatic Change*, Vol. 109, No. 1, August 2011). RCP 6.0, the moderate warming scenario, is characterized by an average global increase of approximately 2.2°C (4°F) by the end of the century, a moderate global effort to curb emissions, and a mix of renewable and nonrenewable energy generation. For more information on the RCPs, readers should refer to van Vuuren et al., 2011, which discusses the scenario development process in detail and is considered the seminal work on the subject.

CHAPTER 2

THE LEVANT AND EGYPT

THE LEVANT AND EGYPT SUBREGION of the CENTCOM AOR consists of Egypt, Iraq, Israel, Jordan, Lebanon, and Syria. Despite its relatively small area compared with the other subregions, it contains a wide variety of climates: arid desert (hot and cold), arid steppe (hot and cold), and warm temperate with dry, hot summers.[1] Most of the land is too arid for agriculture, but the subregion contains multiple rivers and tributaries, including the Nile, Tigris, and Euphrates rivers, which enable farmers to irrigate crops. Although Egypt relies mainly on surface water from the Nile, the country is also on top of the largest fossil aquifer in the world, the Nubian sandstone aquifer system.[2] Jordan and Syria depend heavily on groundwater resources, which are currently being overexploited.[3] Meanwhile, Israel is a global leader in wastewater reclamation and desalination, and Lebanon receives relatively ample precipitation each year.[4]

This section describes the three most notable climate changes that emerged from our analysis for the Levant and Egypt subregion: extreme heat, drought and dryness, and coastal inundation. We also highlight one key environmental impact: the loss of surface water availability.

Extreme Heat

Historically, this subregion has experienced high geographic variability in extreme heat. Lebanon has experienced the lowest number of extreme heat days—about eight days per year above 95°F (35°C) between 1976 and 2005—and Iraq the highest at 137 days per year. Israel (30 days per year historically), Jordan (60 days per year), Syria (77 days per year), and Egypt (102 days per year) fall in between. Iraq, in particular, is one of the hottest places on Earth, and within the past few years, has seen temperatures well above 95°F, including mul-

tiple heat waves exceeding 120°F (50°C) in both Basra and Baghdad.[5]

Looking further out, by 2050, climate projections indicate that extreme heat days will increase throughout the subregion. Lebanon could experience five times more extreme heat days (29 to 40 days per year), Israel could experience three times more (76 to 94 days), and Jordan could record double the number of extreme heat days (100 to 120 days). At the upper end of the range, Egypt and Iraq are not projected to see similar increases in the number of extreme heat days, but they could still experience the most in the subregion at 180 days per year. Syria could experience about one-third of the year with extreme heat by 2050. Figure 2.1 shows the geographic distribution of these changes at subnational scales.

In addition to the effects of temperature alone, rising humidity can exacerbate heat and is prevalent in many of the climates of the subregion. Figure 2.2 shows the number of days of dangerous heat—as measured by the heat index, a combination of relative humidity and temperature—for each country in the subregion in 2050. The heat index measures how heat feels to the human body and combines relative humidity with temperature.[6] When a high temperature and a high relative humidity occur simultaneously, air temperatures can feel much hotter and are more dangerous to human health. A heat index classification of "Danger" indicates that "heat cramps or heat exhaustion [are] likely, and heat stroke [is] possible with prolonged exposure and/or physical activity." A classification of "Extreme Danger" indicates that "heat stroke [is] highly likely."[7] Figure 2.2 shows that Syria is expected to see the highest number of heat index days, close to 30, classified at "Danger" by 2050. Iraq, Jordan, and Lebanon could have up to 20 days or more. Syria and Jordan could see 15 more days per year with a heat index at the "Extreme Danger" classification. Although Egypt has relatively high temperatures, it also

Figure 2.1. Change in Number of Days with Temperatures above 95°F in the Levant and Egypt by 2050

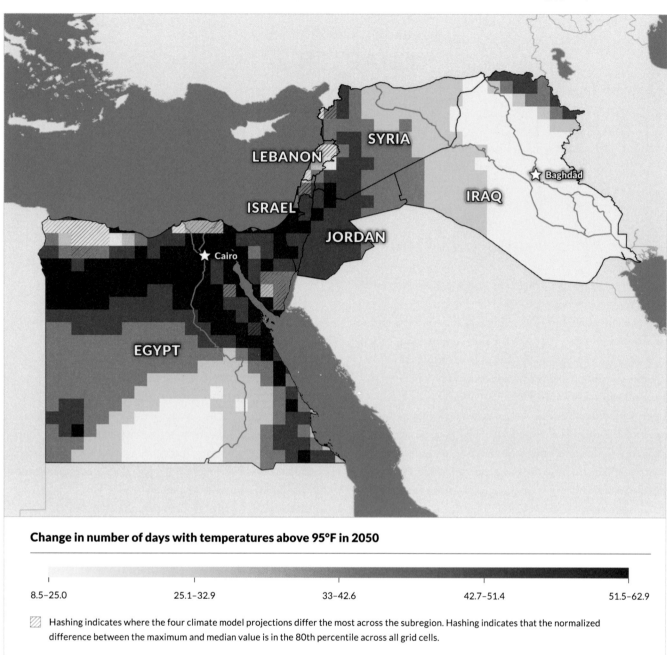

Change in number of days with temperatures above 95°F in 2050

| 8.5–25.0 | 25.1–32.9 | 33–42.6 | 42.7–51.4 | 51.5–62.9 |

Hashing indicates where the four climate model projections differ the most across the subregion. Hashing indicates that the normalized difference between the maximum and median value is in the 80th percentile across all grid cells.

SOURCE: Features authors' estimates from the Inter-Sectoral Impact Model Intercomparison Project (ISIMIP), "ISIMIP2b," database, last published January 31, 2021.

NOTE: Change is in absolute terms as the difference from the historical baseline (1976–2005). Reported values are the median over all climate scenarios and models.

Figure 2.2. Number of Heat Index Days at the "Danger" Classification in the Levant and Egypt in 2050

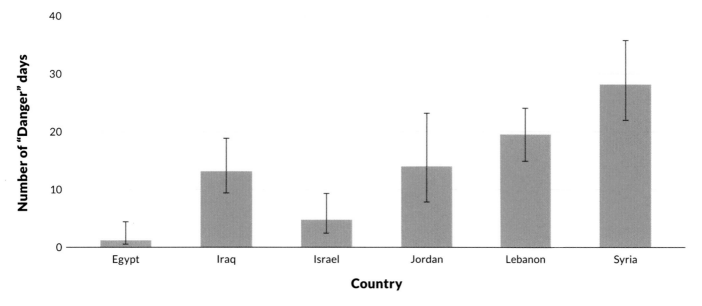

SOURCE: Features authors' estimates from ISIMIP2b data (ISIMIP, 2021).

NOTES: The height of the bars shows the median across climate models, while the error bars show the minimum and maximum across climate models.

has low humidity, which results in a lower heat index value compared with the other countries in the subregion.

Extreme heat has a variety of effects on the environment and society, including health impacts and heat-related mortality, limited safe working hours, electricity outages, and unrest. For Iraq, which has been identified as one of the countries in the AOR most vulnerable to climate change, one study on extreme heat and worker conditions suggests that Iraq has already lost the equivalent hours of 11,000 full-time jobs to extreme heat, and that number could grow to 90,000 by 2030.[8] Heat waves from 2018 to 2022 coincided with a period of political crisis for the country and resulted in power outages, particularly in the southern governorates of Basra, Dhi Qar, and Maysan.[9] Apart from Iraq, extreme heat still poses a significant danger to public health and livelihoods across the subregion. These effects will be more severe in countries with limited capacity to adapt, such as Syria.

Drought and Dryness

Precipitation in the Levant and Egypt subregion is highly variable, ranging from less than 1 inch of annual precipitation in Egypt to almost 30 inches of annual precipitation in Lebanon in the 1976–2005 baseline period. Despite this variability, the region is consistently becoming drier over time. To measure the extent of this trend, we calculated the maximum number of consecutive days in a year without measurable precipitation.[10] This is an important measure to consider when making water management plans, because even small amounts of intermittent precipitation can replenish soil moisture in arid regions. As shown in Figure 2.3, Egypt is the driest country in the Levant and Egypt subregion, with historically around 260 consecutive days without measurable precipitation in a year. This number is projected to increase over time across climate models and climate scenarios. By 2070, in a high-emission scenario, Egypt could experience 318 consecutive days without precipitation. The rest of the subregion sees a similar increasing trend, although some countries, such as Israel and Lebanon, show a smaller increasing trend than others. The median change in Iraq across climate models is an additional 15 days without precipitation.

Aridity is a chronic hazard that, in the Levant and Egypt, is worsening over time.[11] Compared with historical levels, the aridity index is decreasing, meaning that the subregion is becoming drier, as shown in Figure 2.4. Precipitation to potential evapotranspiration (PET) depends on temperature in addition to other climatic variables, such as wind speed, net solar radiation, and vapor pressure deficit. Thus, as temperatures rise, PET increases. At the same time, precipitation is declining in the Levant and Egypt subregion. The cumulative

Figure 2.3. Average Historical and Projected Consecutive Days Without Measurable Precipitation in the Levant and Egypt

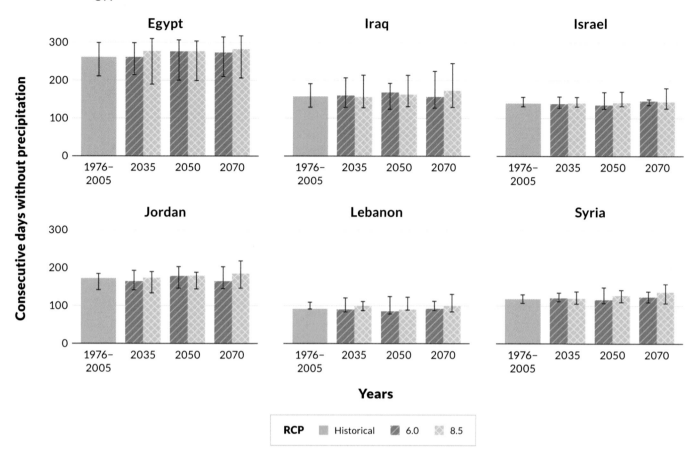

SOURCE: Features authors' estimates from ISIMIP2b data (ISIMIP, 2021).

NOTES: The height of the bars shows the median across climate models, while the error bars show the minimum and maximum across climate models for each RCP.

effect of these processes is a noticeable increase in aridity (i.e., decrease in the aridity index). Aridification increases the risk of soil erosion and therefore the loss of soil nutrients necessary for biodiversity and agriculture.

Coastal Inundation

Within the entire AOR, coastal Egypt and Iraq are projected to see the greatest impacts of sea level rise. Sea level rise and its resulting physical changes to the environmental and human landscape could influence economic trends and social stressors in both countries. Both areas house the outlet of major river basins, creating low-lying, flood-prone areas. In Egypt, the Nile River discharges into the Mediterranean Sea via the Nile Delta, which stretches between Alexandria, Egypt's second largest city, and Port Said. In southern Iraq, the Tigris and

Euphrates River basin enters the Persian Gulf southeast of Basra, Iraq's third largest city. As the home of major population centers that are in close proximity to freshwater sources, coastal Egypt and Iraq are also critical economic centers. In the Nile Delta, agriculture, industry, and fisheries contribute to nearly 20 percent of national gross domestic product (GDP).[12] In Iraq, the Mesopotamian Valley is where barley, dates, rice, and other forms of agriculture are cultivated, and it is the location of the cities of Basra and Nasiriyah, the latter of which is Iraq's fourth largest city.[13] Basra governorate also includes the country's largest oil fields and main port.[14]

When considering the timelines for projected sea level rise, Alexandria could see 0.3 m of sea level rise between 2050 and 2070 but possibly as early as 2040 under more pessimistic climate scenarios.[15] Port Said will see slightly slower rates of sea level rise, with projections showing 0.3 m between 2055 and 2075.[16] Despite these slower rates, coastal inundation in

Figure 2.4. Average Historical and Projected Aridity in the Levant and Egypt

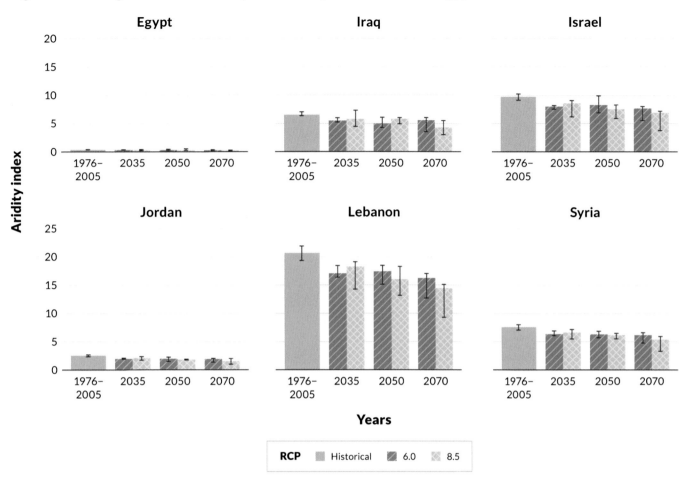

SOURCE: Features authors' estimates from ISIMIP2b data (ISIMIP, 2021).

NOTES: The height of the bars shows the median across climate models, while the error bars show the minimum and maximum across climate models for each RCP. Future projections for Egypt range from 0.26 to 0.35. The historical aridity index is 0.37 for Egypt.

Port Said could be more widespread, as it sits at a lower elevation than Alexandria. Additionally, the entire Nile Delta is experiencing rates of land subsidence that are exaggerating the effect of sea level rise. The northern Nile Delta is estimated to be sinking at a rate of 2.4 mm/year, which is in part the result of a lack of natural sediment deposition that historically came from upstream.[17]

Figure 2.5 shows an analysis of coastal inundation that combines projected sea level rise with expected annual coastal flooding.[18] It shows that large portions of both Port Said and Alexandria could experience coastal flooding by 2050.[19] Other studies estimate that nearly 3,000 square km of the Nile Delta could be permanently inundated by 2100.[20]

Figure 2.5 additionally illustrates projections of coastal inundation in southeastern and coastal Iraq, which will reach further upland into the Mesopotamian Valley by

2050, affecting agricultural and urban areas. The timeline for this encroachment is slightly longer than that for Egypt: Coastal Iraq may not see the same 0.3 m of sea level rise until 2060–2100.[21] Some estimates show that more than one-third of Basra governate, as well as portions of Maysan and Dhi Qar governorates, could be inundated.[22] Forced drainage of Iraq's marshes across the Mesopotamian Valley by Saddam Hussein as retribution against dissenters in the 1990s set the stage for more dramatic effects from sea level rise. Lower freshwater levels in the marshes, made worse by long-term drought, have already allowed the inundation of Persian Gulf seawater inland, destroying farmland.[23]

Given the economic and social importance of these subregions to their respective countries, sea level rise could be highly disruptive and destabilizing. In Egypt, if sea level rise, groundwater pumping, and oil and gas drilling continue at

Figure 2.5. Coastal Inundation Projections in the Levant and Egypt in 2050

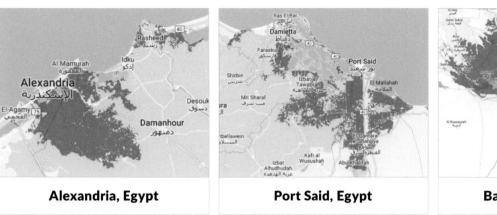

Alexandria, Egypt

Port Said, Egypt

Basra and Coastal Iraq

SOURCE: Features information from Climate Central, undated.

NOTE: The red color indicates projected inundation locations.

current rates, the nearly 3,000 square km of inundated land in the Nile Basin could lead to the displacement of 5.7 million people.[24] Currently, adaptation efforts to stem the inundation in the Nile Delta are slow, exacerbated by a large number of small landowners, who make large-scale, rapid adaptation more difficult.[25] Additionally, rising sea levels have already threatened antiquities and historic preservation efforts in Alexandria and the structural integrity of coastal heritage sites.[26] In Iraq, sea level rise could lead to permanent impacts on Iraq's only port, Umm Qasr, on agriculture in the Mesopotamian Valley, and on the economy of Basra.[27] We found no estimates of the likely number of displaced people for Iraq; however, Basra and Dhi Qar governorates alone were home to more than five million people in 2020.[28] Finally, in both locations, sea level rise can disrupt water supplies by mixing saline seawater with fresh groundwater and surface water. This could affect agriculture, any industries that require fresh water, and residential and public health uses.

Surface Water Availability

All countries in the subregion rely heavily on surface water for residential and agricultural needs, and much of this surface water is shared with neighboring countries, making transboundary water issues sources of cooperation or tension. In Jordan, nearly half of the national water demand comes from crop irrigation, much of it from the Yarmouk River, which is shared between Syria and Jordan.[29] Jordan is considered one of the most water-scarce countries in the world, which means

the country relies on a small amount of renewable water to meet its needs.[30] As a result, all residents regularly face water shortages, and about half receive water only one day per week.[31] Syria faces similar water resource issues, even though it has more available surface water as the home of the Yarmouk, Tigris, and Euphrates rivers, among others. However, Syria shares nearly all surface water sources with neighboring countries. Years of conflict have also damaged domestic and agricultural water infrastructure, leading to water resource shortages even when sufficient water is available.

Many countries in the subregion are lower riparian states, meaning they do not control the flow of a river. Egypt depends heavily on the Nile River for agriculture and livelihoods, particularly in the Nile Delta, and for energy generation at the Aswan Dam. The needs of the Aswan Dam and the construction of the Grand Ethiopian Renaissance Dam (GERD) decrease flows downstream into Egypt and the Nile Delta (see Box 2.1). Israel's and Iraq's surface water resources are affected by being lower riparian states. Iraq is the furthest downstream country of the Tigris and Euphrates rivers, both of which originate in Turkey. Israel receives the majority of its surface water from the Jordan River Basin, but the basin is oversubscribed, with more than 90 percent of its flows diverted by littoral countries.[32] In contrast, Lebanon has a relatively high amount of surface water compared with the rest of the subregion, however, population and water management issues have resulted in water shortages.[33]

Looking forward to 2050, our analysis of hydrologic modeling shows that all countries in this subregion are likely to see

Figure 2.6. Change in Median Surface Water Runoff in the Levant and Egypt in 2050

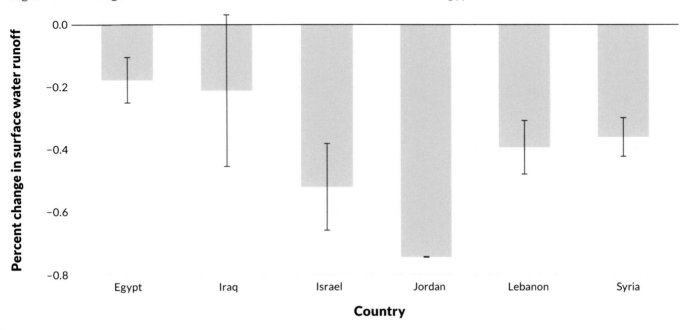

SOURCE: Features authors' estimates from ISIMIP2b data (ISIMIP, 2021).
NOTES: The height of the bars shows the median across climate models, while the error bars show the minimum and maximum across climate models.

a significant loss in surface water by 2050.[34] Figure 2.6 shows the percentage change in surface water runoff between historical conditions and 2050. It shows that Jordan may lose more than 70 percent of its surface water runoff, a significant loss to a country that depends highly on scarce water resources. Israel could see a loss of more than 50 percent of surface water runoff. Reductions in surface water at these scales could lead to significant losses in or changes to agricultural production, further escalating the depletion of groundwater resources and water shortages for domestic, industrial, or commercial users and driving a host of implications for water quality as surface water flows decline. Furthermore, because of the high tensions that already exist in the subregion over shared water resources—most notably concerning the Nile, Jordan, Tigris, and Euphrates river basins—this amount of loss will necessitate enhanced cooperation to maintain the status quo.[35] Of note, any escalating tensions could have crossgeographic combatant command implications. The headwaters of the Nile are located in U.S. Africa Command's AOR, and the headwaters of the Tigris and Euphrates are located in U.S. European Command. Inter-command coordination and strategies will be necessary to defuse issues.

THE LEVANT AND EGYPT

Box 2.1. Water Availability and Tensions Around the Grand Ethiopian Renaissance Dam

Ethiopia and Egypt, along with several other riparian countries, share the waters of the Nile River Basin. The river is prone to both drought and flooding, which makes the addition of reservoirs that can serve as both flood control measures and sources of water during times of drought appealing across the basin. After Egypt was allocated a majority of Nile flows following agreements in 1929 and 1959, Egypt constructed the Aswan Dam to support agriculture and domestic water needs and to generate hydropower. Although Egypt is the most downstream country on the Nile, these early agreements also gave Egypt veto power over the construction of upstream reservoirs that have the potential to disrupt flows to Egypt. However, not all riparian countries were present during the negotiations of the early river basin agreements, including Ethiopia, and those countries have argued that they are not bound by them.[a]

With similar goals as Egypt in the first half of the 20th century, Ethiopia proposed and constructed the GERD. About 85 percent of the Nile's flow into Egypt comes from Ethiopia; early on, Egypt rejected the construction of the GERD, stating that "if it loses one drop, our blood is the alternative" in reference to water loss resulting from the dam's construction and operation.[b] In parallel with the development of the GERD, Ethiopia and other riparian countries signed a water sharing agreement as a part of the Nile Basin Initiative: the Cooperative Framework Agreement (CFA). The CFA did not impose direct downstream flow requirements (i.e., the amount of water that must pass from one country into another), but it did stipulate that signatories should not "significantly affect the water security and current uses and rights of any other Nile Basin State."[c] Although Ethiopia has signed the CFA, Egypt has not and maintains that its water security should not be affected at all by upstream countries. Research has already shown a more than 14 percent loss in the surface area of the Aswan Dam reservoir—Lake Nasser—likely as a result of the filling of the GERD reservoir.[d] This

means that Egyptian water security is highly dependent on GERD reservoir operations and on the upstream climate in Ethiopia.

Projections of Drought in 2035

Ethiopia is very susceptible to drought, as most of its precipitation comes from large-scale weather patterns during a few months of the year. The country typically experiences drought during El Niño years.[e] The impacts of droughts are felt unevenly across the country because of its vast hydroclimatic regimes. These localized climatic dynamics are not well characterized in global climate models, which means they are limited in their ability to project changes in precipitation.

However, we offer a few implications. As the world becomes hotter, there will be increases both in irrigation demand for crops and in drought and dryness. Drought and dryness in the region upstream of the GERD could encourage Ethiopia to release less water downstream to maintain its ability to generate hydropower and ensure water supplies for national needs. Projections indicate that portions of Ethiopia could see up to 300 days per year without precipitation by 2035. The area around the GERD is wetter but depends on flows from regions more susceptible to drought. It is likely that GERD reservoir management will be affected by drought in the future.

[a] Mwangi S. Kimenyi and John Mukum Mbaku, "The Limits of the New 'Nile Agreement,'" Brookings Institution, April 28, 2015.
[b] "Egypt Warns Ethiopia over Nile Dam," Al Jazeera, June 11, 2013.
[c] Kimenyi and Mbaku, 2015.
[d] Mohammed A. El-Shirbeny and Khaled A. Abutaleb, "Monitoring of Water-Level Fluctuation of Lake Nasser Using Altimetry Satellite Data," *Earth Systems and Environment*, Vol. 2, No. 4, May 2018.
[e] Getachew Alem Mera, "Drought and Its Impacts in Ethiopia," *Weather and Climate Extremes*, Vol. 22, October 2018.

Endnotes

1   M. Kottek, J. Grieser, C. Beck, B. Rudolf, and F. Rubel, "World Map of the Köppen-Geiger Climate Classification Updated," webpage, World Maps of Köppen-Geiger Climate Classification, version June 2006. These categories are based on the Köppen-Geiger climate classifications, which is the most frequently used system. These descriptions correspond to the following classifications: BWk, BWh, BSk, BSh, and Csa.

2   Catherine A. Nikiel and Elfatih A. B. Eltahir, "Past and Future Trends of Egypt's Water Consumption and Its Sources," *Nature Communications*, Vol. 12, No. 1, July 2021; Mamdouh Shahin, "The Nubian Sandstone Basin in North Africa, A Source of Irrigation Water for Desert Oases," *Subsurface-Water Hydrology: Proceedings of the International Conference on Hydrology and Water Resources, New Delhi, India, December 1993*, Springer, 1996.

3   Amelia Altz-Stamm, Jordan's *Water Resource Challenges and the Prospects for Sustainability*, term paper for GIS for Water Resources, University of Texas at Austin, Fall 2012; Aden Aw-Hassan, Fadel Rida, Roberto Telleria, and Adriana Bruggeman, "The Impact of Food and Agricultural Policies on Groundwater Use in Syria," *Journal of Hydrology*, Vol. 513, May 2014.

4   Eran Friedler, "Water Reuse—An Integral Part of Water Resources Management: Israel as a Case Study," *Water Policy*, Vol. 3, No. 1, 2001; Erica Spiritos and Clive Lipchin, "Desalination in Israel," in Nir Becker, ed., *Water Policy in Israel: Context, Issues, and Options*, Springer, 2013.

5   Ehab Sh. Ahmed and Ahmed S. Hassan, "The Impact of Extreme Air Temperatures on Characteristics of Iraq Weather," *Iraqi Journal of Science*, Vol. 59, No. 2C, March 2018; Zahraa M. Hassan, Monim H. Al-Jiboori, and Hazima M. Al-Abassi, "The Effect of the Extreme Heat Waves on Mortality Rates in Baghdad During the Period (2004–2018)," *Al-Mustansiriyah Journal of Science*, Vol. 31, No. 2, 2020.

6   National Weather Service, "What Is the Heat Index?" webpage, undated.

7   National Weather Service, undated.

8   United Nations Iraq, "Iraq on Track in the Preparation of Its Climate Change National Adaptation Plan," press release, July 28, 2022; International Labour Organization, *Working on a Warmer Planet: The Impact of Heat Stress on Labour Productivity and Decent Work*, 2019.

9   Louisa Loveluck and Mustafa Salim, "Iraq Broils in Dangerous 120-Degree Heat as Power Grid Shuts Down," *Washington Post*, August 7, 2022; Shawn Yuan, "Heatwaves Scorch Iraq as Protracted Political Crisis Grinds On," Al Jazeera, August 6, 2022. See our second report in this series, *Pathways from Climate Change to Conflict in U.S. Central Command*, which contains a case study of the Basra heat waves and the resulting protests and political crisis (Nathan Chandler, Jeffrey Martini, Karen M. Sudkamp, Maggie Habib, Benjamin J. Sacks, and Zohan Hasan Tariq, *Pathways from Climate Change to Conflict in U.S. Central Command*, RAND Corporation, RR-A2338-2, 2023.).

10 *Measurable precipitation* is defined as equal to or greater than 0.01 inches per day.

11 *Aridity* is defined as the ratio of precipitation to potential evapotranspiration (PET), where PET is the total moisture demanded by the atmosphere. Whereas actual evapotranspiration is the combined amount of moisture that evaporates off the surface and is transpired by plants, potential evapotranspiration is the total amount of moisture that would be evaporated or transpired if the area was limited only by energy and not by the availability of water.

12 Engy Abdel Wahab, "How 'Climate-Smart' Crops Could Prove a Lifeline for Vulnerable Smallholders on the Nile Delta," United Nations Development Programme, May 23, 2022.

13 Columbia University, "Land Use and Agricultural Map of Iraq," webpage, undated. The Mesopotamian Plain is the area that stretches between the Tigris and Euphrates rivers. Here, we focus on the portion southeast of Baghdad.

14 Nahlah Abbas, Sultana Nasrin, Nadhir Al-Ansari, and Sabah H. Ali, "The Impacts of Sea Level Rise on Basrah City, Iraq," *Open Journal of Geology*, Vol. 10, No. 12, December 2020.

15 National Aeronautics and Space Administration (NASA), "Sea Level Projection Tool," webpage, undated.

16 NASA, undated.

17 Esayas Gebremichael, Mohamed Sultan, Richard Becker, Mohamed El Bastawesy, Omar Cherif, and Mustafa Emil, "Assessing Land Deformation and Sea Encroachment in the Nile Delta: A Radar Interferometric and Inundation Modeling Approach," *Journal of Geophysical Research: Solid Earth*, Vol. 123, No. 4, April 2018.

18 Climate Central, "Coastal Risk Screening Tool," interactive map, undated.

19 The data are drawn from Climate Central's "Coastal Risk Screening Tool" (undated) using the median of the Coupled Model Intercomparison Project (CMIP) 6 model output based on the SSP3-7.0 climate scenario. For more information on the methodology, see Scott A. Kulp and Benjamin H. Strauss, "New Elevation Data Triple Estimates of Global Vulnerability to Sea-Level Rise and Coastal Flooding," *Nature Communications*, Vol. 10, No. 1, October 2019. Inundation extents include both permanent and coastal flooding (e.g., tidal).

20 Gebremichael et al., 2018.

21 NASA, undated.

22 Abbas et al., 2020.

23 Wil Crisp, "After Comeback, Southern Iraq's Marshes Are Now Drying Up," Yale Environment 360, January 10, 2023.

24 Gebremichael et al., 2018.

25 Tim McDonnell, "The Nile Delta Isn't Ready for Climate Change," Quartz, October 27, 2022.

26 Salwa Samir, "Extreme Heat Takes Toll on Egypt's Archaeological Heritage," *Al-Monitor*, September 2, 2021.

27 Abbas et al., 2020.

28 Directorate of Population and Manpower Statistics, *Iraq Population Estimates (2020)*, Central Bureau of Statistics Iraq, November 2020.

29 International Trade Administration, *Jordan—Country Commercial Guide*, December 14, 2022.

30 UNICEF Jordan, "Water, Sanitation, and Hygiene," webpage, undated.

31 UNICEF Jordan, undated.

32 Gidon Bromberg, Munqeth Mehyar, and Nader Khateeb, "The Jordan River," Middle East Institute, June 18, 2008.

33 Amin Shaban, "Striking Challenges on Water Resources of Lebanon," in Muhammad Salik Javaid, ed., *Hydrology: The Science of Water*, InTechOpen, 2019.

34 All impact modeling used the same climate data from the GCMs as inputs.

35 United Nations Economic and Social Commission for Western Asia and Bundesanstalt für Geowissenschaften und Rohstoffe, *Inventory of Shared Water Resources in Western Asia*, 2013.

CHAPTER 3

CENTRAL GULF

THE CENTRAL GULF SUBREGION is comprised of Bahrain, Iran, Kuwait, Oman, Qatar, Saudi Arabia, the United Arab Emirates (UAE), and Yemen. It is the most climatologically consistent of the three subregions, a likely result of its location on or proximity to the Arabian Peninsula (excepting Iran),[1] which is characterized as a hot arid desert.[2] Despite Iran's climatic variability, it remains broadly unsuitable for agriculture, along with the vast majority of the subregion. Additionally, in contrast with the Levant and Egypt and Central and South Asia, the Central Gulf does not contain any major rivers, relying almost entirely on groundwater and desalination to meet its water demands.[3] Still, revenues from extensive oil and gas reserves enable most of these dry countries to pump enough groundwater and produce enough desalinated water to consume more water per capita than most regions in the world.[4]

This section describes the three most notable climate changes that emerged from our analysis for the Central Gulf subregion: extreme heat, drought and dryness, and extreme precipitation. We also highlight one key environmental impact: changes in the coastal environment, with a focus on fisheries.

Extreme Heat

The Central Gulf contains some of the hottest countries in the world. As seen in Figure 3.1, Bahrain, Kuwait, Qatar, and the UAE already experience 170–180 days per year with temperatures above 95°F, and this number is expected to increase to more than 200 days per year by 2050.[5] Similar to other locations in the AOR, these countries also regularly experience temperatures above 120°F. All other countries in this subregion will also see increases in days per year with temperatures

above 95°F. Oman and Yemen could see a near doubling in the number of extreme heat days. Oman experienced 60 extreme heat days per year historically and could see close to 115 days by 2050. Yemen is one of the cooler countries in this subregion, but it is still expected to see up to 72 extreme heat days per year by 2050 compared with 38 days historically. Iran, which experienced 76 extreme heat days historically, is projected to see up to 120 days per year of extreme heat by 2050.

When looking at the combined effects of humidity and temperature, projections suggest that Bahrain, Qatar, and the UAE could see more than 50 days per year classified as "Danger" by the heat index and an additional five to eight days per year classified as "Extreme Danger." Although Iran is projected to see around 25 days per year at the heat index "Danger" classification in 2050, it could see an additional 30 days per year at "Extreme Danger," making it one of the countries in the AOR most exposed to conditions that are extremely hazardous to human health.

These types of changes are likely to have serious implications for all countries in this subregion. Because Danger and Extreme Danger heat index conditions pose serious risks to human health, outdoor activities will need to be significantly limited to protect the population. In recent years, nearly all countries in this subregion implemented a ban on outdoor work in the afternoon during the hottest months of the year. Heat-related morbidity and mortality has come to the fore regionally as well, particularly among vulnerable populations required to work outside. During the construction of facilities for the 2022 FIFA World Cup, extreme heat in Qatar was reported as contributing to heat-related deaths of outdoor workers.[6] Moreover, research has shown that extreme heat—including associated health, well-being, and productivity

Figure 3.1. Average Historical and Projected Number of Extreme Heat Days in the Central Gulf

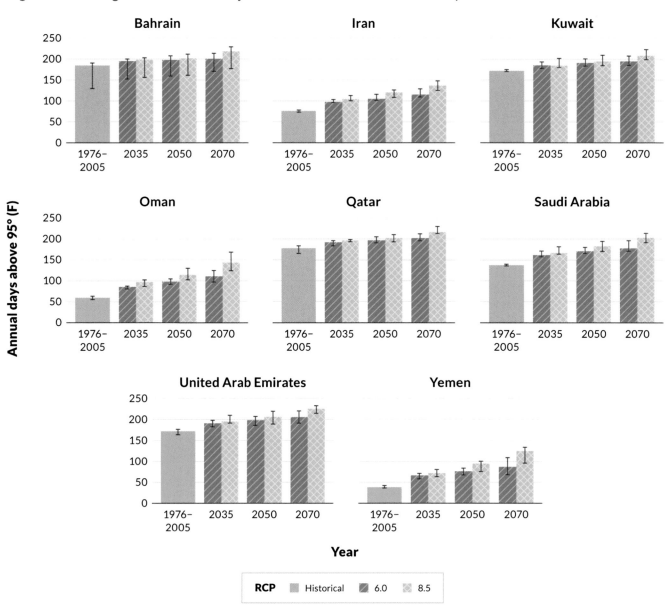

SOURCE: Features authors' estimates from ISIMIP2b data (ISIMIP, 2021).

NOTES: The height of the bars shows the median across climate models, while the error bars show the minimum and maximum across climate models for each RCP.

effects—has already contributed to GDP losses in the subregion. In 1995, Qatar, Bahrain, and the UAE are estimated to have lost nearly 2 percent of GDP on average because of heat stress.[7] This same study further projects additional GDP losses to increase by 2030; at the top of that list is Qatar, with the potential to lose more than 3 percent of GDP, even with moderate global temperature increases.[8]

Drought and Dryness

With the majority of the subregion classified as a warm desert climate, this area is exceptionally dry, receiving a historical average of 3.83 inches of precipitation per year. This scarcity of fresh water has led to a reliance on both groundwater and desalination to meet national water needs. Oil-rich countries, such as Kuwait, Qatar, Saudi Arabia, and the UAE, can use their relative wealth and proximity to the ocean to desalinate vast amounts of seawater. Indeed, Kuwait, Qatar, and the UAE rely on desalination for more than 90 percent of their drinking water needs.[9]

Additionally, the subregion is projected to get drier over time because of climate change. Figure 3.2 shows the change in consecutive days without measurable precipitation relative to the historical baseline. Northern Iran in particular will experience a noticeable increase in dry days, between 6 and 37, as will western Saudi Arabia. (See Box 3.1. for more information about water resource management in Iran.)

Extreme Precipitation

With a consistent, warm desert climate across the subregion, the Central Gulf is climatically dry on average. However, some areas are prone to intense precipitation and thus are at risk of flooding. The hot and dry conditions of the subregion throughout most of the year induce a relatively impermeable land surface. This occurs in part because hotter temperatures increase the amount of moisture demanded by the atmosphere, increasing the amount of evaporation from the land surface and drying the soil. Over time, this drying creates a crust-like surface on the land, increasing the likelihood of precipitation running off the surface. This runoff can lead to flash flooding. In 2022, almost every country in the subregion had some portion of its territory affected by flash flooding after days of heavy precipitation.

Figure 3.3 shows the maximum amount of precipitation that falls within 24 hours in each period across the Central Gulf countries. According to our modeled results, extreme precipitation is projected to worsen relative to baseline in Iran, Oman, Saudi Arabia, the UAE, and Yemen. If the region does not adapt to these trends, flash flooding and the associated safety hazards and damage to infrastructure will worsen as well. Such adaptations could include building more robust infrastructure, improving stormwater systems, and moving infrastructure and communities away from floodplains.

The relatively rapid warming of the Arabian Sea has contributed to an increase in the frequency and severity of tropical cyclones.[10] Tropical cyclones can occur in the months before and after the monsoon, which occurs from June to September annually. Rains from the Indian monsoon do not affect countries in the Central Gulf as heavily as in South Asia, but Iran, Oman, and Yemen can still experience heavier rains because of monsoonal dynamics.

Box 3.1. Water Resource Management in Iran

Iran receives a historical average of 10.1 annual inches of precipitation. Although this is 2.5 times as much as the subregional average, the country is still susceptible to droughts and has experienced a long-term loss in national water resources. As shown in Figure 3.3, the number of days per year without precipitation is projected to increase as a result of climate change. This increase is especially prominent near major population centers, such as Tehran. Although climate change will worsen water scarcity, the main drivers of water scarcity in Iran are population growth, inefficient agricultural practices, and the mismanagement of water resources.[a]

In Iran, 93 percent of all water use is by the agricultural sector, and farmers are especially affected by the occurrence of drought.[b] In 2021, a severe drought accompanied by extreme heat led to rolling blackouts and insufficient water for agriculture. These events resulted in protests that turned deadly.[c]

Furthermore, because of increasing water demand because of population growth, important water bodies, such as Lake Urmia and Lake Hamun, are shrinking as dams are built to store water in reservoirs.[d] Groundwater is also overexploited. Overpumping of groundwater wells is so severe that the Tehran Plain is estimated to be experiencing the highest rates of land subsidence in the world.[e] Recently, as a mitigation measure, the government has begun increasing its production of desalinated water because of worsening droughts.[f]

[a] Kaveh Madani, "Water Management in Iran: What Is Causing the Looming Crisis?" *Journal of Environmental Studies and Sciences*, Vol. 4, No. 4, December 2014.
[b] National Research Council, *Water Conservation, Reuse, and Recycling: Proceedings of an Iranian-American Workshop*, National Academies Press, 2005.
[c] Kareem Fahim and Miriam Berger, "Protests over Water Shortages in Iran Turn Deadly in a Summer of Drought and Rolling Blackouts," *Washington Post*, July 21, 2021.
[d] Madani, 2014.
[e] Madani, 2014. Groundwater pumping from certain soil types can lead to soil compaction as it dries, causing subsidence—or the lowering—of the land surface.
[f] Banafsheh Keynoush, "With the Hope Line, Iran Aims to Boost Seawater Transfer to Fight Growing Drought," Middle East Institute, June 9, 2021.

Figure 3.2. Change in the Number of Consecutive Dry Days in the Central Gulf by 2050

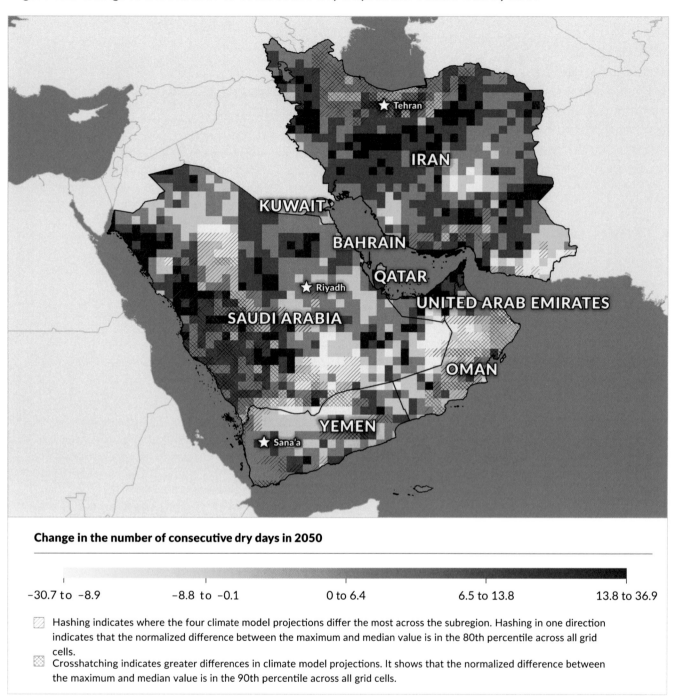

Change in the number of consecutive dry days in 2050

| −30.7 to −8.9 | −8.8 to −0.1 | 0 to 6.4 | 6.5 to 13.8 | 13.8 to 36.9 |

Hashing indicates where the four climate model projections differ the most across the subregion. Hashing in one direction indicates that the normalized difference between the maximum and median value is in the 80th percentile across all grid cells.

Crosshatching indicates greater differences in climate model projections. It shows that the normalized difference between the maximum and median value is in the 90th percentile across all grid cells.

SOURCE: Features authors' estimates from ISIMIP, 2021.

NOTE: Change is in absolute terms as the difference from the historical baseline (1976–2005). Reported values are the median over all climate scenarios and models.

Figure 3.3. Maximum 24-Hour Precipitation in the Central Gulf

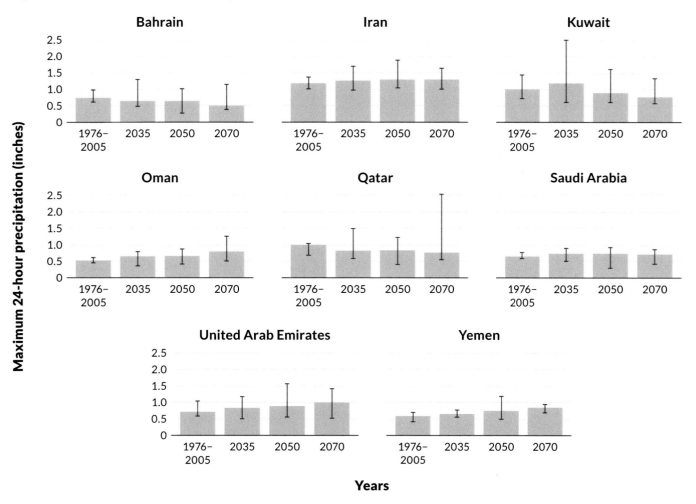

SOURCE: Features authors' estimates from ISIMIP2b data (ISIMIP, 2021).

NOTES: The height of the bars shows the median across climate models, while the error bars show the minimum and maximum across climate models.

Fisheries

Although the oil and gas industry dominates economies across the Central Gulf, the aquaculture and fishing industry is a critical element of the economies of certain countries. Before civil war began in 2015, Yemen's fishing sector was its second largest revenue-producing sector after oil and gas.[11] The aquaculture industry in Iran employs more than 200,000 people.[12] Surveys administered in Oman also indicate an interest among younger generations in pursuing employment in aquaculture.[13] The fisheries sector in Oman is growing rapidly, recording a growth rate of 44 percent in 2020.[14]

However, a combination of climate change and local environmental policies have contributed to decreasing fish stocks

in the Central Gulf subregion. Our results using a marine ecosystems model, the Object-oriented Simulator of Marine ecOSystEms (OSMOSE) (see Figure 3.4), show the greatest impact on Oman and Yemen, with a clear, declining trend in total consumer biomass through 2070. Rising ocean temperatures likely contribute to this decline. The relatively rapid warming of the sea surface has led to a noticeable reduction in dissolved oxygen levels in the Arabian Sea, which negatively affects aquaculture.[15]

In the Persian Gulf, environmental and economic policies also drive changes in total fish catch. For example, countries such as Bahrain, Qatar, and the UAE are increasing their

Figure 3.4. Average Total Fish Catch

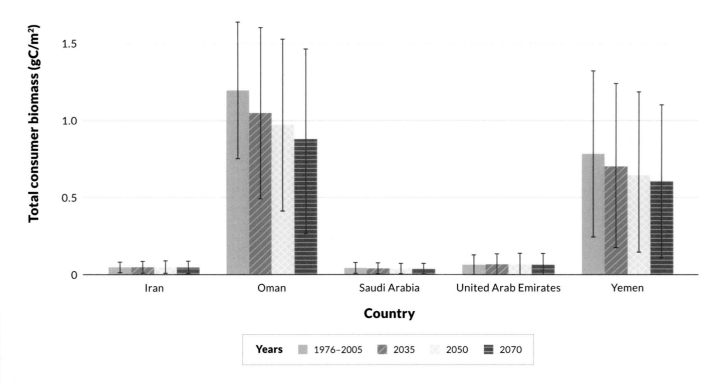

SOURCE: Features authors' estimates from ISIMIP2b data (ISIMIP, 2021).

NOTES: NOTE: gC/m² = gas chromatography per square meter. The height of the bars shows the median across climate models, and the error bars show the minimum and maximum across climate models.

production and dependence on desalination plants to provide fresh water to their populations.[16] The effluent brine from desalination plants has a negative impact on the habitat of marine life.[17] When combined with commercial overfishing trends, it is likely that aquaculture decline will continue, affecting economic and social stability, if adaptive and mitigative actions are not taken.[18] These decreases in supply are occurring in concert with rising aquaculture demand. (See Box 3.2 for an analysis of the localized impacts of sea level rise in the Central Gulf.) In Iran, representatives from the Iran Fisheries Organization have announced a target of doubling aquaculture production by 2025.[19]

Box 3.2. Sea Level Rise in the Central Gulf

By 2050, the Central Gulf will begin to see localized impacts from sea level rise. On the basis of our analysis of data on coastal inundation that combines projected sea level rise with expected annual coastal flooding, the following areas will be at the greatest risk from sea level rise:[a]

- Bahrain: Bahrain International Airport, coastal Muharraq governorate, and Southern governorate
- Kuwait: coastline of Kuwait City and Bubiyan Island
- Qatar: Hamad International Airport and southeastern Qatar
- Saudi Arabia: Dammam
- UAE: Abu Dhabi barrier islands.

[a] Climate Central, undated.

Endnotes

[1] Iran encompasses seven climate zones, ranging from hot arid desert (BWh) to snow with dry summers (Dsb and Dsc).

[2] BWh in the Köppen-Geiger climate classification nomenclature (Kottek et al., 2006).

[3] Tarek Ben Hassen and Hamid El Bilali, "Chapter 26—Water Management in the Gulf Cooperation Council: Challenges and Prospects," *Current Directions in Water Scarcity Research*, Vol. 5, 2022.

[4] Esmat Zaidan, Mohammad Al-Saidi, and Suzanne H. Hammad, "Sustainable Development in the Arab World—Is the Gulf Cooperation Council (GCC) Region Fit for the Challenge?" *Development in Practice*, Vol. 29, No. 5, July 2019.

[5] Compared with our analysis of modeled temperature from 1976 to 2005.

[6] International Labour Organization, *One Is Too Many: The Collection and Analysis of Data on Occupational Injuries in Qatar*, November 2021.

[7] International Labour Organization, 2019.

[8] International Labour Organization, 2019.

[9] Najmedin Meshkati, "Gulf Escalation Threatens Drinking Water," Belfer Center for Science and International Affairs, Harvard Kennedy School, June 26, 2019.

[10] Majid Pourkerman, Nick Marriner, Sedigheh Amjadi, Razyeh Lak, Mohammadali Hamzeh, Gholamreza Mohammadpor, Hamid Lahijani, Morteza Tavakoli, Christophe Morhange, and Majid Shah-Hosseini, "The Impacts of Persian Gulf Water and Ocean-Atmosphere Interactions on Tropical Cyclone Intensification in the Arabian Sea," *Marine Pollution Bulletin*, Vol. 188, March 2023.

[11] U.S. Agency for International Development, *The Fisheries Sector in Yemen: Status and Opportunities*, November 2019.

[12] Afshin Adeli, "An Analysis of Employment and Fisheries Businesses Opportunities in Iran and the World," *Journal of Utilization and Cultivation of Aquatics*, Vol. 9, No. 1, April 2020. This represents a little

under 1 percent of Iran's total labor force, according to data from the World Bank (World Bank, "Labor Force, Total—Iran, Islamic Rep.," database, 2022).

[13] Iffat S. Chaudhry, Asma H. Al-Harthi, Ghada M. Al-Shihimi, Khalsa M. Al-Saiti, and Maryam S. Al-Siyabi, "Millennials Outlook Towards Opportunities and Challenges in Fishing Industry of Muscat Governorate of Oman," *Middle East Journal of Management*, Vol. 4, No. 1, 2017.

[14] "Fisheries Sector Sees Growth of 44.5% in 2020," *Oman Daily Observer*, November 20, 2021.

[15] Zouhair Lachkar, Michael Mehari, Muchamad Al Azhar, Marina Lévy, and Shafer Smith, "Fast Local Warming Is the Main Driver of Recent Deoxygenation in the Northern Arabian Sea," *Biogeosciences*, Vol. 18, No. 20, October 2021.

[16] Francesco Paparella, Daniele D'Agostino, and John A. Burt, "Long-Term, Basin-Scale Salinity Impacts from Desalination in the Arabian/Persian Gulf," *Scientific Reports*, Vol. 12, November 2022.

[17] W. J. F. Le Quesne, L. Fernand, T. S. Ali, O. Andres, M. Antonpoulou, J. A. Burt, W. W. Dougherty, P. J. Edson, J. El Kharraz, J. Glavan, R. J. Mamiit, K. D. Reid, A. Sajwani, and D. Sheahan, "Is the Development of Desalination Compatible with Sustainable Development of the Arabian Gulf?" *Marine Pollution Bulletin*, Vol. 173, Part A, December 2021.

[18] Technological mitigation measures include using diffusers to dilute the brine, deep well injection, and evaporation ponds (Le Quesne et al., 2021). However, these measures have accompanying risks: Brine diffusers limit salinity locally but not regionally, deep well injection may cause groundwater contamination, and evaporation ponds can cause the death of wildlife that mistake the pond as a viable habitat.

[19] "Iran Targets Doubling Aquaculture Production by 2025," *Tehran Times*, November 16, 2022.

CHAPTER 4

CENTRAL AND SOUTH ASIA

CENTRAL AND SOUTH ASIA is the most geographically expansive of the three subregions, spanning more than 30 degrees of latitude or approximately 2,200 miles—roughly the distance between Seattle and New Orleans. Because of this wide area, the climatological characteristics of this subregion are also highly variable. The countries in this subregion are Afghanistan, Kazakhstan, Kyrgyzstan, Pakistan, Tajikistan, Turkmenistan, and Uzbekistan. Despite geographic features that vary from coastal regions to extensive mountain ranges to steppes, the subregion is generally arid. This arid climate precludes extensive farming, yet intensive irrigation near the main rivers in the subregion, the Syr Darya and Amu Darya in Central Asia and the Indus in Pakistan, enables the production of crops, such as cotton and wheat.[1]

This chapter describes the three most notable climate changes that emerged from our analysis of the Central and South Asia subregion: extreme heat, extreme cold, and dust storms. We also highlight one key environmental impact: changes in crop production.

Extreme Heat and Extreme Cold

Although temperatures are warming across Central and South Asia, this is the only CENTCOM subregion to experience both extreme heat and extreme cold regularly. The annual average temperatures for each country range from as low as 34°F in Kyrgyzstan to 68°F in Pakistan. Tajikistan is projected to have up to 20 extreme heat days per year by 2050, compared with eight days historically. Turkmenistan experienced 62 extreme heat days per year historically and is expected to see up to 106 days by 2050. Pakistan is the hottest country in this subregion, with projections showing extreme heat of 141 days per year by 2050 (105 days per year historically). Kyrgyzstan is the coolest, with no days above 95ºF in the historical data and one projected per year by 2050. Afghanistan falls in between; projections for Afghanistan show up to 141 extreme heat days per year by 2050 (54 days per year historically). Figure 4.1 shows the spatial distribution of these changes.

Although other subregions in the AOR had more days with extreme heat, days at or above 95ºF may be more impactful in this subregion, which has historically had lower temperatures, signaling the likely need for greater adaptation. In Pakistan, for example, a 2022 heat wave contributed to crop losses of up to 30 percent in the Punjab region.[2] Research suggests that climate change has made heat waves, such as the ones between March and May 2022, nearly 30 times more likely in successive decades.[3] Heat events in 2021 also affected public health, leading to increased rates of heat-related mortality and morbidity in Turkmenistan and Tajikistan.[4]

Extreme cold, which is unique to this AOR subregion, is also lessening. Countries experiencing extreme cold events may see 20–30 fewer days below freezing per year by 2050 (i.e., frost days), with Kyrgyzstan and Tajikistan expected to see the largest reductions in days below freezing. Less extreme cold could lengthen the growing season for this subregion. In isolation from other climate impacts, this could be a positive change, but continued drought is likely to dampen this benefit.

Figure 4.1. Change in Number of Days with Temperatures Above 95°F in Central and South Asia by 2050

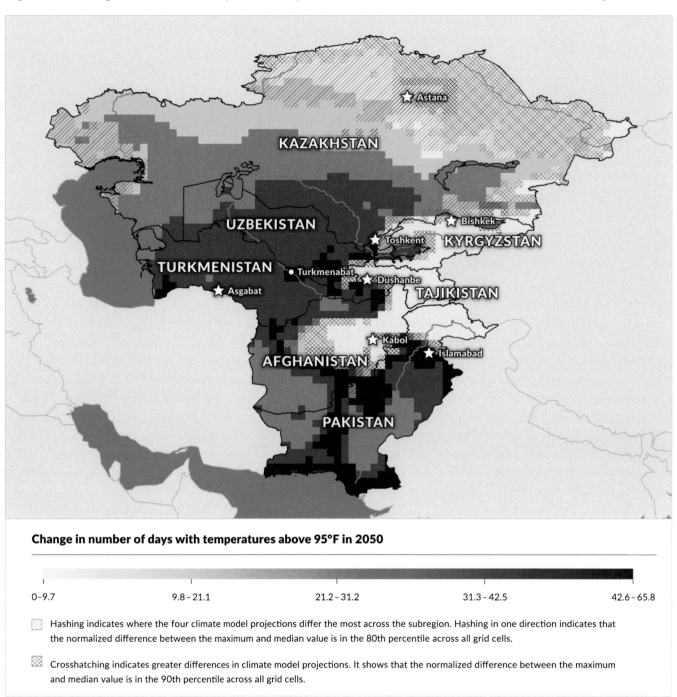

Change in number of days with temperatures above 95°F in 2050

| 0-9.7 | 9.8-21.1 | 21.2-31.2 | 31.3-42.5 | 42.6-65.8 |

Hashing indicates where the four climate model projections differ the most across the subregion. Hashing in one direction indicates that the normalized difference between the maximum and median value is in the 80th percentile across all grid cells.

Crosshatching indicates greater differences in climate model projections. It shows that the normalized difference between the maximum and median value is in the 90th percentile across all grid cells.

SOURCE: Features authors' estimates from ISIMIP2b data (ISIMIP, 2021).

NOTES: Change is in absolute terms as the difference from the historical baseline (1976–2005). Reported values are the median over all climate scenarios and models.

Dust Storms

Dust storms are relatively common occurrences in this subregion, affecting public health, agricultural production, and air transport, among other impacts.[5] In particular, research has shown that more than 80 percent of the population in Tajikistan, Turkmenistan, and Uzbekistan is exposed to unhealthy air quality from dust storms.[6] The deposition of dust following dust storms is also detrimental to agriculture in these three countries, and high salt levels in regional dust storms can damage crops.[7] In Kazakhstan, dust storms are more common in the southern and more arid regions of the country.[8] Across the subregion, areas of high agricultural or economic activity are more prone to dust storms because of the higher availability of sediments that could become airborne.

Looking forward to 2050, two variables that can influence the incidence and severity of dust storms are daily wind speeds and aridity. When wind speeds cross the threshold at which dust can become airborne (approximately 7.7 meters/second), and aridity remains high or worsens, climate-induced increases in dust storms could occur.[9] Figure 4.2 shows that the number of days capable of producing dust storms will stay relatively constant for the subregion, with only modest declines for Kazakhstan and Uzbekistan. At the same time, aridity is likely to stay fairly high for the subregion into the future. These analyses suggest that climate change alone will likely not lead to large-scale increases in dust storms, although continued long-term aridity combined with poor land management practices could increase the number of dust storms in the future.

Crop Production

Not unlike in the Central Gulf, a combination of the effects of climate change and environmental policies could have an outsize impact on the subregion's economic productivity. Despite the limited amount of suitable land because of aridity, agriculture is a critical sector in Central and South Asia. The amount of arable land in relation to total land varies from 40 percent in Pakistan to as low as 4 percent in Turkmenistan.[10] Agriculture as a percentage of GDP varies from 5 percent in Kazakhstan to 27 percent in Tajikistan.[11] Historically, the subregion as a whole receives an average of 11.7 inches of precipitation annually, and country averages range from 6.6 inches of precipitation annually in Turkmenistan to 17.4 inches annually in Kyrgyzstan.

With limited annual rainfall, the subregion therefore relies heavily on irrigation for crop production and specifically on surface water rather than groundwater. The reliance on surface water puts Central Asia at high risk for crop failures from climate change. As the Syr Darya and Amu Darya rivers are fed primarily by snow and glacier melt, which are highly susceptible to rising temperatures, irrigation requirements could be put at risk.[12] In 2021, Central Asia experienced an extreme agricultural drought that led to massive losses in crops and livestock.[13] Research indicates that the decreasing trend in soil moisture in Central Asia is likely to continue.[14] See Box 4.1 for an example of how heavier than average rainfall in Pakistan can affect ground surfaces.

Other than the risks from water scarcity, climate change could affect the subregion's crop production through changes in temperature and carbon dioxide (CO_2) concentrations. Increasing temperature in the subregion both reduces the incidence of frost days and increases the risk of extreme heat days. Although general temperature increases could lengthen the growing season in Central Asia, extreme heat endangers crops in critical stages in their growth. These risks offset the benefits from increases in average temperature.[15] Rising temperatures also increase potential evapotranspiration, which decreases soil moisture over time and further reduces the amount of water available for use by crops in an already arid environment. Additionally, increasing concentrations of CO_2 in the atmosphere will also affect crop production with the CO_2 fertilization effect. Research has shown that the increase of CO_2 concentration in the atmosphere leads to an increase in photosynthesis, thus supporting overall plant growth globally.[16] The overall effect of how yields will change is highly uncertain, but research has shown that irrigation requirements per unit of crop production will decrease as a result of this effect.[17]

Agriculture is a large contributor to the economy of Central and South Asia. Thus, changes in the subregion's agricultural output can have a substantial effect on the quality of life and overall stability of the subregion. We analyzed the change in maize yields in 2050 relative to the historical baseline. Maize yields, rather than cotton or wheat yields, were considered because of their availability in the Community Land Model (CLM) version 4.5 in the ISIMIP database.[18] Maize is therefore intended to be a representative crop for other staple crops.[19] Figure 4.3 shows the relationship between the change in maize yields in 2050 relative to the historical baseline and latitude in Central and South Asia. In general, maize yields exhibit greater increases as latitude increases. This trend is supported by other agricultural models in the literature.[20]

Figure 4.2. Historical and Projected Number of Windy Days in Central and South Asia in 2050

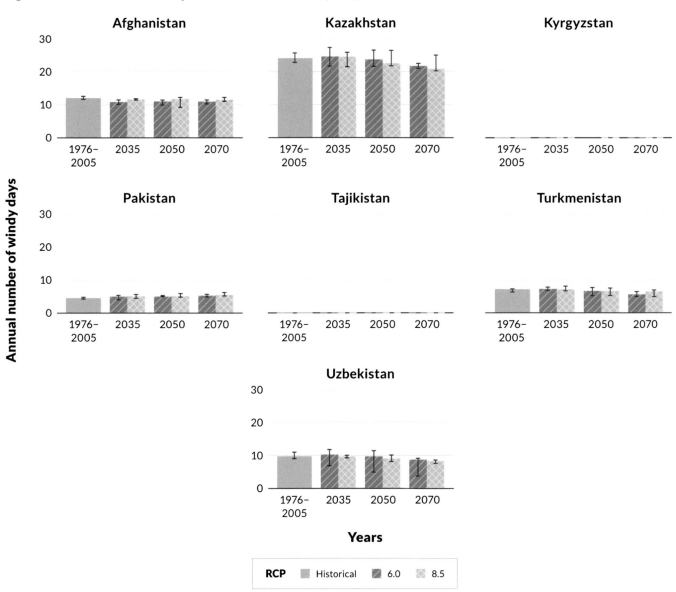

SOURCE: Features authors' estimates from ISIMIP2b data (ISIMIP, 2021).

NOTES: The height of the bars shows the median across climate models, while the error bars show the minimum and maximum across climate models for each RCP. For Kyrgyzstan and Tajikistan, all models reported zero or near zero windy days (i.e., annual number of days with surface wind speeds above 7.7 m/s).

Figure 4.3. Change in Maize Yields in Central and South Asia in 2050

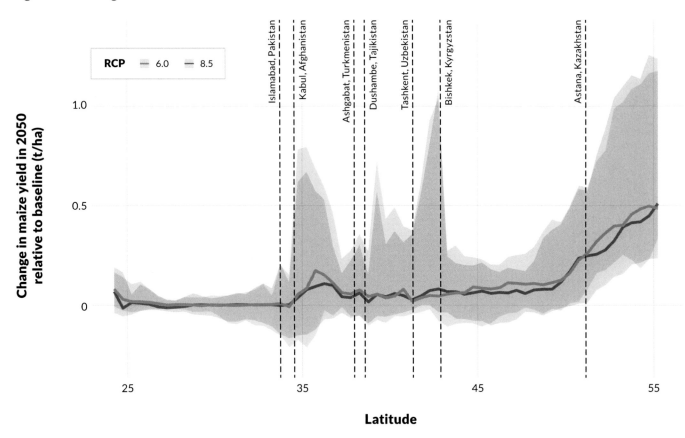

SOURCE: Features authors' estimates from ISIMIP2b data (ISIMIP, 2021).

NOTES: t/ha = tons per hectare. The two RCPs are shown as different colors. The lines depict the median of all grid cells of a certain latitude among the four climate models within each RCP. The ribbons show the range between the 5th and 95th percentiles of all grid cells of a certain latitude. Dashed lines show the latitudes of each capital in the subregion.

Box 4.1. Flooding in Pakistan

From June to October 2022, Pakistan was hit by a series of devastating floods that led to billions of dollars in damage and thousands of deaths. The floods were the result of heavier than usual monsoonal rains. South Asia experiences heavier rains during La Niña years, and 2022 was the third consecutive La Niña. Because of climate change, the monsoon in Pakistan is projected to begin later and bring heavier rainfall. The delayed start will extend the dry season and may lead to a more impermeable surface that will exacerbate the risk of flash flooding when the monsoon begins. Coastal areas of Pakistan are also at risk of flooding from the projected increase in the severity and frequency of tropical cyclones in the Arabian Sea.

Pakistan experienced floods of a similar magnitude in 2010 because of heavier-than-usual rainfall during a La Niña. After the resulting devastation, the government constructed 22 "disaster-resilient" villages that were intended to resist future flooding. Clearly, these adaptive measures were insufficient to prevent the devastation from the 2022 floods. Given the increasing risks of damage from flooding in the coming decades, there is an urgent need to repair and build infrastructure that can withstand floods of similar and heightened magnitude as those in 2022.

Projections of Extreme Precipitation in Future Years

As part of our analysis of climate change in the AOR, we calculated the maximum amount of precipitation that fell in a 24-hour period. This metric gives an indication of the intensity of rainfall,

which directly contributes to the risk of flooding. We found that, on average throughout the country, extreme precipitation will increase over time compared with the historical baseline (see Figure 4.4). The upper bound of these estimates in 2070 is a 67 percent increase from the upper bound of the historical data.

Figure 4.4. Maximum 24-Hour Precipitation for Pakistan in 2050

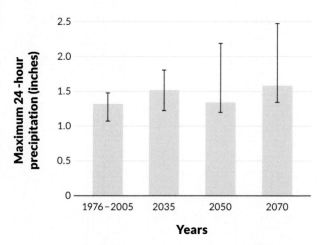

SOURCE: Features authors' estimates from ISIMIP2b data (ISIMIP, 2021).

NOTES: The height of the bars shows the median across climate models, while the error bars show the minimum and maximum across climate models.

Endnotes

1 M. M. Aldaya, G. Muñoz, and A. Y. Hoekstra, *Water Footprint of Cotton, Wheat and Rice Production in Central Asia*, UNESCO-IHE Institute for Water Education, March 2010.

2 "Climate Change Made Heatwaves in India and Pakistan '30 Times More Likely,'" World Meteorological Organization, May 24, 2022.

3 "Climate Change Made Heatwaves," 2022.

4 Central Asian Bureau for Analytical Reporting, "Abnormal Heat in Central Asia: Society Must Take This Problem Seriously," 2021.

5 Asian and Pacific Centre for the Development of Disaster Information Management, *Sand and Dust Storms Risk Assessment in Asia and the Pacific*, August 2021.

6 Asian and Pacific Centre for the Development of Disaster Information Management, 2021.

7 Asian and Pacific Centre for the Development of Disaster Information Management, 2021.

8 Gulnura Issanova, Jilili Abuduwaili, Azamat Kaldybayev, Oleg Semenov, and Tatiana Dedova, "Dust Storms in Kazakhstan: Frequency and Division," *Journal of the Geological Society of India*, Vol. 85, March 2015.

9 Pattanun Achakulwisut, Susan C. Anenberg, James E. Neumann, Stefani L. Penn, Natalie Weiss, Allison Crimmins, Neal Fann, Jeremy Martinich, Henry Roman, and Loretta J. Mickley, "Effects of Increasing Aridity on Ambient Dust and Public Health in the U.S. Southwest Under Climate Change," *GeoHealth*, Vol. 3, No. 5, 2019.

10 World Bank, "Arable Land (% of Land Area)," database, undated.

11 U.S. Agency for International Development, 2019.

12 Yue Qin, John T. Abatzoglou, Stefan Siebert, Laurie S. Huning, Amir AghaKouchak, Justin S. Mankin, Chaopeng Hong, Dan Tong, Steven J. Davis, and Nathaniel D. Mueller, "Agricultural Risks from Changing Snowmelt," *Nature Climate Change*, Vol. 10, No. 5, April 2020.

13 There are three main types of drought: meteorological, hydrologic, and agricultural. Meteorological drought is characterized by below-average levels of precipitation, hydrologic drought occurs from below-average surface water runoff, and agricultural drought occurs because of below-average soil moisture. Note that a discussion of socioeconomic drought is outside the scope of this report.

14 Jie Jiang and Tianjun Zhou, "Agricultural Drought over Water-Scarce Central Asia Aggravated by Internal Climate Variability," *Nature Geoscience*, Vol. 16, No. 2, 2023.

15 Yang Liu, Xiu Geng, Zhixin Hao, and Jingyun Zheng, "Changes in Climate Extremes in Central Asia Under 1.5 and 2 °C Global Warming and Their Impacts on Agricultural Productions," *Atmosphere*, Vol. 11, No. 10, 2020.

16 Zaichun Zhu, Shilong Piao, Ranga B. Myneni, Mengtian Huang, Zhenzhong Zeng, Josep G. Canadell, Philippe Ciais, Stephen Sitch, Pierre Friedlingstein, Almut Arneth, Chunxiang Cao, Lei Cheng, Etsushi Kato, Charles Koven, Yue Li, Xu Lian, Yongwen Liu, Ronggao Liu, Jiafu Mao, Yaozhong Pan, Shushi Peng, Josep Peñuelas, Benjamin Poulter, Thomas A. M. Pugh, Benjamin D. Stocker, Nicolas Viovy, Xuhui Wang, Yingping Wang, Zhiqiang Xiao, Hui Yang, Sönke Zaehle, and Ning Zeng, "Greening of the Earth and Its Drivers," *Nature Climate Change*, Vol. 6, No. 8, April 2016.

17 Jing Tian and Yongqiang Zhang, "Detecting Changes in Irrigation Water Requirement in Central Asia Under CO2 Fertilization and Land Use Changes," *Journal of Hydrology*, Vol. 583, April 2020.

18 ISIMIP, 2021.

19 According to the Food and Agriculture Organization, the water requirement for maize is 500–800 mm per growing period. Cotton requires 700–1300 mm, and wheat requires 450–650 mm of water per growing period (Food and Agriculture Organization, *Irrigation Water Management: Irrigation Water Needs*, Training Manual No. 3, 1986).

20 Jonas Jägermeyr, Christoph Müller, Alex C. Ruane, Joshua Elliott, Juraj Balkovic, Oscar Castillo, Babacar Faye, Ian Foster, Christian Folberth, James A. Franke, Kathrin Fuchs, Jose R. Guarin, Jens Heinke, Gerrit Hoogenboom, Toshichika Iizumi, Atul K. Jain, David Kelly, Nikolay Khabarov, Stefan Lange, Tzu-Shun Lin, Wenfeng Liu, Oleksandr Mialyk, Sara Minoli, Elisabeth J. Moyer, Masashi Okada, Meridel Phillips, Cheryl Porter, Sam S. Rabin, Clemens Scheer, Julia M. Schneider, Joep F. Schyns, Rastislav Skalsky, Andrew Smerald, Tommaso Stella, Haynes Stephens, Heidi Webber, Florian Zabel, and Cynthia Rosenzweig, "Climate Impacts on Global Agriculture Emerge Earlier in New Generation of Climate and Crop Models," *Nature Food*, Vol. 2, No. 11, November 2021.

CENTRAL AND SOUTH ASIA

CHAPTER 5

—

CONCLUSIONS

THIS STUDY CENTERS on characterizing the relationships between climate change and conflict to inform operational and longer-term decisionmaking by CENTCOM. In this report, we focused our analysis on conducting a regional climate assessment that quantifies changes in key climate hazards—extreme heat, extreme cold, drought and dryness, extreme precipitation, dust storms, and sea level rise—and their effects on food and water security. To do this, we analyzed climate model outputs to characterize how these climate hazards are projected to change across the CENTCOM AOR in 2035, 2050, and 2070. We also used a global land surface model and a global hydrologic model to study how changes in climate hazards could affect crop production and water availability, as well as a marine ecosystems model to examine changes to fisheries.

Summary of Findings

Across the AOR, the overwhelming trend is toward a hotter and drier climate. This is evidenced in the chronic trends of precipitation and temperature and in their extremes (e.g., extreme heat). Changes are most noticeable at the extremes, as days above a hazard threshold increase in frequency and hazards overall become more severe. The co-occurring nature of trends in extreme heat and dryness mean that the corresponding impacts are compounded. Extreme heat increases the demand for water, while drought can exacerbate heat by reducing the land's ability to cool itself.

Table 5.1 groups countries by their primary climate hazards of concern, identified as the hazards with the largest percentage of changes relative to baseline. Nearly all countries in the AOR are increasingly exposed to extreme heat, and almost half of the countries are facing significant decreases in water avail-

ability alongside heightened extreme heat. Two countries—Bahrain and Kuwait—are facing loss of water availability more acutely than other hazards. These countries will still see extreme heat in the future, but increases in temperatures may not be significantly different from current levels. Hotter and drier conditions are likely to mean increased water stress for much of the region, particularly for those countries already at heightened rates of water scarcity, such as Iraq, Jordan, and Syria. Oman, Tajikistan, and Yemen are additionally exposed to extreme precipitation. Appendix B provides the data underlying this assessment.

As the AOR becomes drier, countries will put increasing pressure on existing water resources. Without adequate governance over these resources, such pressures may become contributing factors to conflict. This could be particularly acute across the AOR compared with other parts of the world, given existing water scarcity issues and the high degree of shared water resources among nations already prone to conflict. Depending on the water source, conflicts can occur within national boundaries or internationally, as major rivers, lakes, and aquifers are generally shared by multiple nations. In situations in which there is a lack of trust between the parties that share a resource, individual nations may overexploit the resource in an attempt to use it before their neighbors.[1] Such behavior can be mitigated to some extent by water-sharing agreements, although these are exceedingly difficult to monitor and enforce.

Paradoxically, the drying trend in the AOR also leaves many countries increasingly susceptible to flash flooding when extreme precipitation does occur. The drying of the land surface enables precipitation to run off rather than infiltrate the soil. Such countries as Yemen, Oman, and Pakistan are especially at risk of more severe flash flooding in the future

Table 5.1. Hazards of Primary Concern by Country

Main Hazards	Countries
Water availability	• Bahrain • Kuwait
Extreme heat and water availability	• Afghanistan • Egypt[a] • Iran • Iraq[a] • Israel • Jordan • Lebanon • Qatar • Syria • Turkmenistan
Extreme heat	• Kazakhstan • Kyrgyzstan • Pakistan • Saudi Arabia • United Arab Emirates • Uzbekistan
Extreme heat and extreme precipitation	• Oman • Tajikistan • Yemen

[a] Iraq and Egypt will additionally face the risk of significant coastal inundation because of sea level rise.

CONCLUSIONS

because of the jointly occurring hazards of aridification and increases in extreme precipitation.

There are a few key hot spots that will see additional compounding hazards. Southern Iraq—Basra, Maysan, and Dhi Qar governorates—is vulnerable to sea level rise, surface water losses, and extreme heat. Additionally, Alexandria and Port Said in Egypt face risks from sea level rise and declines in surface water availability from the Nile. For both countries, these changes can exacerbate challenges to agriculture and urban areas.

Caveats and Limitations

There are a few key caveats to this analysis. We generally examined climate change in isolation of other effects. For example, on the basis of our analysis of wind speeds and aridity alone, we do not expect a significant increase in dust storms for most of the AOR. However, it is more likely that the con-

tribution of climate change to dust storms may be minor compared with land management or other resource practices, as these factors already contribute significantly to the incidence and severity of dust storms. Our water resource analyses were limited in that the global hydrologic model we used did not consider the effects of glacial retreat on base flows. For this reason, we do not report our findings for the Indus, Syr Darya, or Amu Darya rivers. We also do not highlight groundwater in this report, as there were limitations in our ability to relate net groundwater abstractions to long-term changes in groundwater storage. Finally, because the majority of the analysis in this report was performed at the country level, many subregional and local changes are not depicted. This spatial aggregation reduces the extremes that could be seen with a higher level of resolution.

The findings in this report reflect the trends across climate models and scenarios. Although we did not discuss the uncertainty of the projections at length in the text, this is depicted

32

in the visualizations of the reported metrics either as error bars in the bar plots or as hashing in the maps. The magnitude of uncertainty can have enormous implications for planning. Depending on the level of decisionmaker risk aversion and available resources, a planner may want to make decisions on the basis of the median or worst-case outcome. High levels of uncertainty may also act as a deterrent to act until more information is available. However, it is critical that uncertainty be a part of the decisionmaking process.

Future Work

This report provides an overview of climate projections and the key climate hazards that could affect the CENTCOM AOR through 2050. Through our analysis, we identified key areas for future research that could build on the foundations that this report established. The first of these areas is focused on the next steps in the causal pathways from climate to conflict. Future work could move beyond climate hazards and examine the secondary or tertiary impacts of climate change on people and their well-being and on governments and/or economies. Such an analysis would need to consider the impacts of climate change on critical infrastructure systems, such as water or electricity provision. This would aid in understanding the impacts of climate change on human systems and the implications of these impacts on societal instability. The Jordan Water Model is an example of a complex systems model that is capable of examining the impacts of climate change and other stressors on Jordan's freshwater security.[2] Such work could be extended or serve as a model for future work. Another avenue for this line of work could focus on identifying the thresholds at which climate change could trigger significant regional resource implications (e.g., a large-scale water shortage) that would be destabilizing for society.

A second area for future work relates to climate models and information. This report used projects from the fifth report of the CMIP for the availability of data across the hazards of interest, but more recent international reports (e.g., the IPCC Sixth Assessment Report) rely on more recent data from the next CMIP, CMIP6. Future work could update this analysis to use CMIP6 output, as sufficient data is available. Additionally, as the U.S. Department of Defense (DoD) Climate Assessment Tool (DCAT) is updated in the future, future work could incorporate data from DCAT alongside or in place of CMIP6 data from ISIMIP. Furthermore, we considered only one global hydrologic model in this analysis. Given the important of water to the region and the risks climate poses to it, examining multiple hydrologic models to understand the full scope of plausible changes in water resources would be valuable.

Finally, a third area for future work could be focused specifically on water-related conflict. Given existing tensions over shared water resources in the AOR and the findings of this analysis, which show potential for growing water scarcity, a deeper AOR-wide dive into the implications of climate change on water-related conflict could be conducted. Although some analyses exist for each river basin, a comprehensive regional analysis that uses a standardized set of assumptions, data, and models could enable a more complete and comparable understanding of risks to water systems. This type of work could be used to identify the tipping points for and pathways to climate change-induced, water-related conflict. Additionally, such an analysis could include a characterization of climate change implications on existing water governance, sharing agreements, or the lack thereof.

Other Reports in This Series

As noted previously, this report is the first of a series that investigates the potential impacts of climate change on the security environment in the CENTCOM AOR. This project report addressed the first research task: characterizing the physical environment in the region through 2070 and highlighting those climate hazards that could most significantly affect regional populations and contribute to conflict. To accomplish this, we analyzed climate data and conducted a qualitative literature review to determine which climate hazards we should prioritize for the Middle East and Central Asia. This climatic and environmental overview provides a foundation for the reports that will follow. The next report in this series uses the climate hazards identified in this analysis to investigate potential paths to conflict at three levels: civil disturbances, intrastate conflict, and interstate conflict.[3] Regional case studies will supplement this analysis to illustrate how climate hazards have already contributed to instability and conflict in the AOR. The third report will combine the climate projections with socioeconomic scenarios and historical conflict data to project the potential for violence at the intrastate level within the AOR.[4] For our fourth report, the climate projections and conflict pathways will inform scenarios that could realistically lead to conflict within the AOR and present difficult challenges for the United States and its adversaries: China, Russia, and Iran.[5] In our final report, we present how CENTCOM could use operations, activities, and investments in the coming decades to address security threats related to climate stress, either to mitigate the risk of conflict related to climate hazards or to respond to climate-related conflict via military intervention, such as stabilization operations or humanitarian assistance and disaster relief.[6]

Endnotes

[1] This is an example of the tragedy of the commons.

[2] Jim Yoon, Christian Klassert, Philip Selby, Thibaut Lachaut, Stephen Knox, Nicolas Avisse, Julien Harou, Amaury Tilmant, Bernd Klauer, Daanish Mustafa, Katja Sigel, Samer Talozi, Erik Gawel, Josue Medellín-Azuara, Bushra Bataineh, Hua Zhang, and Steven M. Gorelick, "A Coupled Human–Natural System Analysis of Freshwater Security Under Climate and Population Change," *Proceedings of the National Academy of Sciences*, Vol. 118, No. 14, April 2021.

[3] Chandler et al., 2023.

[4] Mark Toukan, Stephen Watts, Emily Allendorf, Jeffrey Martini, Karen M. Sudkamp, Nathan Chandler, and Maggie Habib, *Conflict Projections in U.S. Central Command: Incorporating Climate Change*, RAND Corporation, RR-A2338-3, 2023.

[5] Howard J. Shatz, Karen M. Sudkamp, Jeffrey Martini, Mohammad Ahmadi, Derek Grossman, and Kotryna Jukneviciute, *Mischief, Malevolence, or Indifference? How Competitors and Adversaries Could Exploit Climate-Related Conflict in the U.S. Central Command Area of Responsibility*, RAND Corporation, RR-A2338-4, 2023.

[6] Karen M. Sudkamp, Elisa Yoshiara, Jeffrey Martini, Mohammad Ahmadi, Matthew Kubasak, Alexander Noyes, Alexandra Stark, Zohan Hasan Tariq, Ryan Haberman, and Erik E. Mueller, *Defense Planning Implications of Climate Change for U.S. Central Command*, RAND Corporation, RR-A2338-5, 2023.

APPENDIX A

SUPPLEMENTAL INFORMATION ON METHODS

Literature Review

To refine our selection of climate hazards relevant to the CENTCOM AOR, we conducted a literature review to understand which variables prior studies have considered and what primary and secondary impacts have been discussed. We looked at gray literature focused on information about what is happening in the region and written for a decisionmaking audience, such as national and regional reports, assessments, and climate change strategies. We also reviewed relevant academic literature that detailed more technical work with climate models and their outputs. From both sets of literature, our goal was to develop a consensus around the climate hazards that are most relevant to the region and the most appropriate metrics to quantify those hazards.

The review process began with reports from the Sixth Assessment Cycle of the IPCC, including the first two working group reports and the IPCC Working Group I Interactive Atlas.[1] On determining that the AOR-defined region does not directly fit into the IPCC-defined regions, we drew content from overlapping regions in the Interactive Atlas and collected mentions of areas within the AOR-defined region throughout relevant chapters and crosschapter papers. We then reviewed sources from important regional and multinational organizations, including the United Nations' Regional Initiative for the Assessment of Climate Change Impacts on Water Resources and Socio-Economic Vulnerability in the Arab Region, the World Bank Group, and the National Security, Military, and Intelligence Panel on Climate Change of the Center for Climate and Security.[2] Finally, we reviewed the DCAT and conducted a review of relevant DoD documents.[3] Although the available information in the DCAT was not appropriate for this study, we did align hazards and metrics as appropriate with DCAT indicators (e.g., use of the same heat index equations, threshold for measurable precipitation, frost days).

From these resources, we were able to extract a full list of climate hazards of significance to the AOR. This list included those hazards that were mentioned across the above resources and were relevant to much of the region. For example, tropical cyclones were relevant to Pakistan but not the rest of the Central and South Asia subregion. Wildfire was also not widely discussed in the literature across any of the subregions. Table A.1 provides the full list of hazards we assessed, along with their metrics and definitions.

Data Sources

The impacts of climate change are highly uncertain, especially decades into the future. This uncertainty stems from the unknown trajectory of future emissions and uncertainty in our understanding of the climate system. To ensure both sources of uncertainty were represented in our analysis, we used the output from four climate models from the fifth iteration of CMIP that were based on two RCPs, plausible emissions trajectories developed by the climate modeling community for uncertainty analysis.[4] For our analysis, we used the high emissions scenario (RCP 8.5) and a moderate warming scenario (RCP 6.0).[5] These scenarios are meant to show what climate changes may occur in the absence of stringent global climate policy (both scenarios) and to present a more pessimistic future for military planners to consider (RCP 8.5). The blue and green color trends in Figure A.1 illustrate how these two scenarios influence average temperature in the AOR. The ribbons within each color depict the structural uncertainty from using different climate models. The four climate models served as input to the environmental impact models.

Comparing outputs across different models and scenarios requires that available output variables and spatial and temporal resolution are consistent. We therefore relied on datasets from the ISIMIP. The ISIMIP contains a subset of datasets from GCMs within CMIP5 and datasets from impact models (e.g., hydrologic and agricultural models). CMIP5 provides a framework to compare outputs of GCMs that were run with the same input datasets (i.e., the same RCP), thereby enabling comparison across models. CMIP simulations are coordinated to align with the IPCC assessments, so that the assessment reports communicate the state-of-the-art in climate modeling. Within the ISIMIP, daily CMIP5 data are available on a 0.5-degree by 0.5-degree grid (i.e., approximately 55 km by 55 km at the equator) from four GCMs (GFDL, HadGEM2, IPSL, and MIROC) for both RCPs.[6]

Within the ISIMIP archive, CMIP5 datasets are used as inputs to environmental systems models, such as hydrologic and agricultural models. To evaluate the changes in surface water and groundwater pumping and recharge, we used output from the global WaterGAP model. WaterGAP contains a global hydrology model that estimates canopy, snow, soil, groundwater, local lakes, local wetlands, global lakes, global reservoirs, and rivers but does not consider glacier storage or melt. For the surface water runoff estimates contained in this report, the model calculates runoff from the soil water balance. Specifically, soil water storage over time is calculated as effective precipitation minus runoff from land and actual evapotranspiration. Runoff from land is then partitioned into surface runoff and groundwater recharge. It also contains a water use model that accounts for irrigation, livestock, domestic, manufacturing, and thermal power uses.[7]

Additionally, we used the CLM4.5 within the ISIMIP archive to model the change in agricultural yield because of changes in precipitation, temperature, and CO_2 concentration. CLM is a commonly used land surface model that contains detailed representations of biophysical processes.[8] We used the OSMOSE model to measure changes in fish stocks because of changes in ocean acidity and temperature.[9] As with the data from GCMs, all impact model data are available on a 55km-by-55km grid. All data sources are shown in Table A.2.

The four GCMs—GFDL, HadGEM, MIROC, and IPSL—were selected from 36 total GCMs included in CMIP5 for their readiness for inclusion in the ISIMIP project. Other work has shown that this subset covers more of the total uncertainty in temperature and precipitation than other randomly sampled GCM subsets of similar size.[10] Although there are other global datasets of temperature and precipitation with higher resolution that use CMIP5 GCMs (e.g., the NASA NEX-GDDP dataset), we chose the ISIMIP dataset because it included a fuller range of output hydroclimatic variables (e.g., wind speeds and PET) in addition to the output of impact models that may have important links to conflict, such as water resources, agriculture, and fish catch. Furthermore, because most of our analysis is conducted at the country scale, downscaling to a finer spatial resolution is not necessary.

Data Processing and Synthesis

Our analytical process generally followed five main steps, as illustrated in Figure A.2. We first downloaded the data from the ISIMIP archive, stored it centrally, and then cropped the global data to the bounds of the CENTCOM AOR.[11] We then used the annual data for each variable (e.g., daily maximum temperature) to perform any necessary calculations to compute the metrics of interest for each grid cell (e.g., annual number of days above 95°F). All metrics are detailed in Table A.1. Using shapefiles of the countries in the AOR and shapefiles of exclusive economic zones (EEZs), we calculated a spatially weighted average of each metric across all cells in each country polygon or EEZ (values in each cell were weighted by their fractional area inside the polygon). For coastal processes, we used EEZs to connect any variables that extend beyond land borders to each coastal country. This process produces daily-to-annual metrics for each climate hazard at the country level. With annual data from 1976 to 2099 for each metric and country, we then temporally averaged to a historical period

Table A.1. Climate Hazards Examined for the U.S. Central Command Area of Responsibility

Climate Hazards	Metrics	Description
Acute Hazards		
Extreme temperature	Annual number of days above 95°F	Counts of the number of days each year with a maximum temperature above 95°F
	Heat index	Uses relative humidity and temperature to calculate the heat perceived by the human body
	Five-day maximum or minimum temperature	Annual maximum or minimum average five-day temperature
	Frost days	Annual number of days with a temperature at or below 32°F
Drought	SPEI[a]	Uses PET and precipitation to characterize drought tendency
	Annual number of days without measurable precipitation	Annual number of days with precipitation less than 0.01 inches
	Consecutive number of dry days	Annual maximum number of consecutive days with precipitation less than 0.01 inches
	Aridity	Ratio of precipitation to PET; provides an indication of how the moisture demanded by the atmosphere relates to the available supply
Dust storms	Surface wind	Annual number of days with surface wind speeds above 7.7 meters/second
Extreme precipitation	Maximum 24-hour precipitation	Highest amount of 24-hour precipitation in a year; provides a measure of precipitation intensity for assessing flood risk
	Maximum five-day precipitation	Highest amount of precipitation in a rolling sum of five days in a year
Chronic Hazards		
Sea level rise	Sea level rise	Total sea level rise, including contributions from glacial and ice sheet melting and thermal expansion
Precipitation trends	Monthly total precipitation	Sum of daily precipitation in a month
	Annual total precipitation	Sum of daily precipitation in a year
Temperature trends	Annual mean temperature	Annual average of daily mean temperature
	Monthly maximum temperature	Maximum temperature in each month within a year
	Monthly minimum temperature	Minimum temperature in each month within a year
Impacts on Environmental Systems		
Water resources	Groundwater change	Net abstractions from groundwater
	Surface water availability	Change in total surface water runoff calculated as the sum of surface and subsurface runoff
Agriculture	Crop production	Crop yield (maize, soy)
	Fisheries	Change in total consumer biomass

NOTE: [a]Santiago Beguería, Borja Latorre, Fergus Reig, and Sergio M. Vicente-Serrano, "About the SPEI," webpage, undated.

and three future periods. The years in the historical period (1976–2005) are averaged to obtain a single number for the baseline. We chose future years that represented the near term (2035), medium term (2050), and long term (2070) with respect to future climate. To ensure that these years captured the overall trends of each metric over time, we calculated the ten-year rolling average of future years before filtering the years 2035, 2050, and 2070.[12] This process was repeated for each climate model and for both climate scenarios, resulting in eight future projections for each metric of interest for 2035, 2050, and 2070.

Figure A.1. Average Temperature in the U.S. Central Command Area of Responsibility

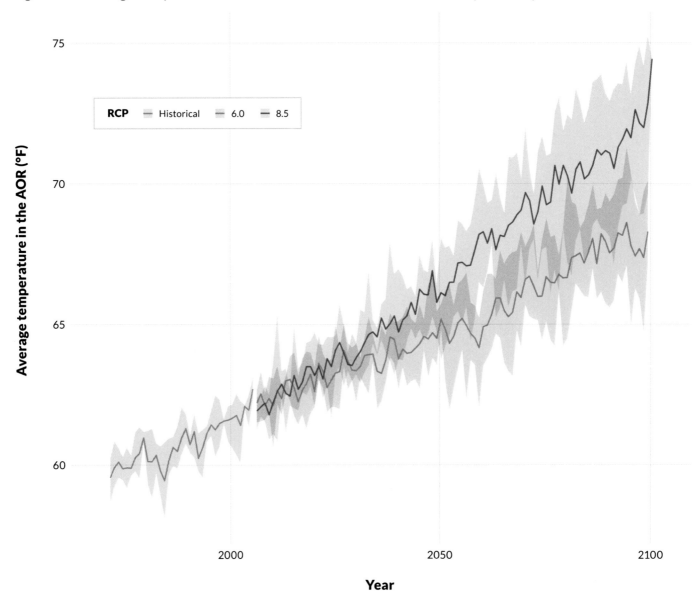

SOURCE: Features authors' estimates from ISIMIP2b data (ISIMIP, 2021).

Table A.2. Characteristics of Data Sources

Dataset	Variables	Period	Temporal Resolution	Spatial Resolution
CMIP5 (GFDL, IPSL, HadGEM2, MIROC5)	• Temperature • Precipitation • Humidity • Surface wind speed	Historical: 1976–2005 Projected: 2006–2099	Daily	~55km grid
WaterGAP	• Groundwater abstractions • Surface water runoff	Historical: 1976–2005 Projected: 2006–2099	Monthly	~55km grid
CLM	• Crop yield	Historical: 1976–2005 Projected: 2006–2099	Annual	~55km grid
OSMOSE	• Total consumer biomass	Historical: 1976–2005 Projected: 2006–2099	Monthly	~55km grid

NOTES: ~ = approximately. WaterGAP runs at a daily temporal resolution but is aggregated to monthly scales in the ISIMIP database to remain consistent with other impact models.

Figure A.2. Overview of Analysis Process

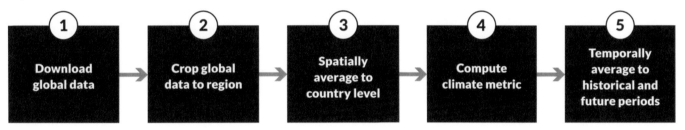

1. Download global data → 2. Crop global data to region → 3. Spatially average to country level → 4. Compute climate metric → 5. Temporally average to historical and future periods

APPENDIX A

Endnotes

1. Valérie Masson-Delmotte, Panmao Zhai, Anna Pirani, Sarah L. Connors, Clotilde Péan, Yang Chen, Leah Goldfarb, Melissa I. Gomis, J. B. Robin Matthews, Sophie Berger, Mengtian Huang, Ozge Yelekçi, Rong Yu, Baiquan Zhou, Elisabeth Lonnoy, Thomas K. Maycock, Tim Waterfield, Katherine Leitzell, and Nada Caud, eds., *Climate Change 2021: The Physical Science Basis. Contribution of Working Group I to the Sixth Assessment Report of the Intergovernmental Panel on Climate Change*, Intergovernmental Panel on Climate Change, Cambridge University Press, 2021; Hans-Otto Pörtner, Debra C. Roberts, Melinda M. B. Tignor, Elvira Poloczanska, Katja Mintenbeck, Andrés Alegría, Marlies Craig, Stefanie Langsdorf, Sina Löschke, Vincent Möller, Andrew Okem, and Bardhyl Rama, eds., *Climate Change 2022: Impacts, Adaptation, and Vulnerability. Contribution of Working Group II to the Sixth Assessment Report of the Intergovernmental Panel on Climate Change*, Intergovernmental Panel on Climate Change, Cambridge University Press, 2022; IPCC, "IPCC WGI Interactive Atlas," webpage, undated; Maialen Iturbide, Jesús Fernández, José Manuel Gutiérrez, Joaquín Bedia, Ezequiel Cimadevilla, Javier Díez-Sierra, Rodrigo Manzanas, Ana Casanueva, Jorge Baño-Medina, Josipa Milovac, Sixto Herrera, Antonio S. Cofiño, Daniel San Martín, Markel García-Díez, Mathias Hauser, David Huard, and Özge Yelekçi, "Repository Supporting the Implementation of FAIR Principles in the IPCC-WG1 Atlas," Zenodo, database, version v2.0-final, August 9, 2021.

2. Regional Initiative for the Assessment of Climate Change Impacts on Water Resources and Socio-Economic Vulnerability in the Arab Region, *Arab Climate Change Assessment Report*, United Nations Economic and Social Commission for Western Asia, 2017; Swedish Meteorological and Hydrological Institute and United Nations Economic and Social Commission for Western Asia, *Future Climate Projections for the Mashreq Region: Summary Outcomes*, Regional

Initiative for the Assessment of the Impact of Climate Change on Water Resources and Socio-Economic Vulnerability in the Arab Region, 2021; World Bank Group, *Egypt Country Climate and Development Report*, CCDR Series, 2022a; World Bank Group, *Iraq Country Climate and Development Report*, CCDR Series, 2022b; World Bank Group, *Jordan Country Climate and Development Report*, CCDR Series, 2022c; World Bank Group, *Kazakhstan Country Climate and Development Report*, CCDR Series, 2022d; World Bank Group, *Pakistan Country Climate and Development Report*, CCDR Series, 2022e; National Security, Military, and Intelligence Panel on Climate Change, *A Security Threat Assessment of Global Climate Change: How Likely Warming Scenarios Indicate a Catastrophic Security Future*, Center for Climate and Security, February 2020.

[3] A. O. Pinson, K. D. White, E. E. Ritchie, H. M. Conners, and J. R. Arnold, *DoD Installation Exposure to Climate Change at Home and Abroad*, U.S. Army Corps of Engineers, April 2021; Department of the Army, Office of the Assistant Secretary of the Army for Installations, Energy and Environment, *United States Army Climate Strategy*, February 2022; Arto Hirvela, *Effects of Climate Change to Balance of Power in the AOR*, Near East South Asia Center for Strategic Studies, July 13, 2021.

[4] At the time of our data processing, CMIP6 climate variables were available in the ISIMIP database, but the environmental systems models (e.g., land surface and hydrologic models) that use CMIP6 inputs were not. For consistent comparison between climate models and environmental systems models, we used CMIP5. Unlike CMIP6 scenarios, CMIP5 climate scenarios are not tied to socioeconomic scenarios. Model runs that required anthropogenic inputs, such as demand, fertilizer levels, and land use change, used the 2005 levels of these variables throughout the scenario (i.e., through 2100). Hydroclimatic variables (e.g., temperature, precipitation, PET, and wind speed) do not require this input data.

[5] To represent the high emissions scenario, we used RCP 8.5. RCP 8.5 is characterized by an average global warming of approximately 3.7°C (6.7°F) by the end of the century, a low global effort to curb emissions, and primarily nonrenewable sources of energy generation at the global scale (van Vuuren et al., 2011). RCP 6.0, the moderate warming scenario, is characterized by an average global increase of approximately 2.2°C (4°F) by the end of the century, a moderate global effort to curb emissions, and a mix of renewable and nonrenewable energy

generation. For more information on the RCPs, readers should refer to van Vuuren et al., 2011. This work discusses the scenario development process in detail and is considered the seminal work on the subject.

[6] The ISIMIP simulation protocol includes validating these datasets using reanalysis data and applying bias correction factors where necessary.

[7] Hannes Müller Schmied, Denise Cáceres, Stephanie Eisner, Martina Flörke, Claudia Herbert, Christoph Niemann, Thedini Asali Peiris, Eklavyya Popat, Felix Theodor Portmann, Robert Reinecke, Maike Schumacher, Somayeh Shadkam, Camelia-Eliza Telteu, Tim Trautmann, and Petra Döll, "The Global Water Resources and Use Model WaterGAP v2.2d: Model Description and Evaluation," *Geoscientific Model Development*, Vol. 14, No. 2, February 2021.

[8] Keith Oleson, David M. Lawrence, Gordon B. Bonan, B. A. Drewniak, Maoyi Huang, Charles D. Koven, Samuel Levis, Fang Li, William J. Riley, Zachary M. Subin, Sean Swenson, Peter E. Thornton, Anil Bozbiyik, Rosie Fisher, Colette L. Heald, Erik Kluzek, Jean-Francois Lamarque, Peter J. Lawrence, L. Ruby Leung, William Lipscomb, Stefan P. Muszala, Daniel M. Ricciuto, William J. Sacks, Ying Sun, Jinyun Tang, and Zong-Liang Yang, *Technical Description of Version 4.5 of the Community Land Model (CLM)*, NCAR Technical Note No. NCAR/TN-503+ STR, National Center for Atmospheric Research, July 2013.

[9] Yunne-Jai Shin and Philippe Cury, "Exploring Fish Community Dynamics Through Size-Dependent Trophic Interactions Using a Spatialized Individual-Based Model," *Aquatic Living Resources*, Vol. 14, No. 2, March 2001.

[10] Carol F. McSweeney and Richard G. Jones, "How Representative Is the Spread of Climate Projections from the 5 CMIP5 GCMs Used in ISI-MIP?" *Climate Services*, Vol. 1, March 2016.

[11] The ISIMIP raw data are in netcdf format, a common file format used to house geospatial data with multiple layers.

[12] This means that each continuous ten-year period (e.g., 2006–2015, 2007–2016, etc.) was averaged so that each year of data represents the average of that year, the four years preceding it, and the five following it. For instance, the data in the year 2050 are the average of the years 2046–2055.

CLIMATE HAZARDS BY COUNTRY

TABLE B.1. SHOWS the country-level percentage changes in 2050 compared with a 1976–2005 historical baseline for extreme heat (number of days above 95°F), water availability (surface water runoff, consecutive dry days), and extreme precipitation (maximum 24-hour precipitation).

Table B.2 shows the country-level values for 2050 and from 1976 to 2005 (shown in brackets).

Table B.1. Percentage Change in Climate Hazards in 2050 by Country

Country	Number of Days Above 95°F	Consecutive Dry Days	Surface Water Runoff (inches/year)	Maximum 24-Hour Precipitation (inches)
Afghanistan	39.7	5.5	−11.4	9.3
Bahrain	7.6	0.0	−23.1	−14.2
Egypt	39.6	5.8	−7.4	−2.2
Iran	47.2	7.7	−13.7	11.2
Iraq	19.0	3.2	−15.5	4.0
Israel	162.6	−0.5	−22.0	−0.4
Jordan	77.0	4.0	−37.1	−15.3
Kazakhstan	160.7	2.1	17.0	12.4
Kuwait	12.4	1.7	−38.4	−15.9
Kyrgyzstan	854.8	8.3	2.1	8.3
Lebanon	289.5	−2.7	−20.4	2.3
Oman	72.8	−8.8	41.3	22.1
Pakistan	30.1	0.3	5.2	1.8
Qatar	11.9	4.4	−1.1	−10.3
Saudi Arabia	28.9	1.5	12.0	12.9
Syria	44.9	1.3	−16.2	−4.4
Tajikistan	120.5	10.6	9.5	14.8
Turkmenistan	57.7	8.8	−11.7	10.2
UAE	17.6	7.9	17.2	11.7
Uzbekistan	84.1	7.9	−5.5	10.1
Yemen	110.2	4.6	31.8	16.6

Percent Difference:

−14.3   >100

NOTE: The percentage difference is calculated on the basis of a historical period of 1976–2005.

Table B.2. Climate Hazards in 2050 and 1976–2005 by Country

Country	Number of Days Above 95°F	Consecutive Dry Days	Surface Water Runoff (inches/year)	Maximum 24-Hour Precipitation (inches)
Afghanistan	75.7 [54.2]	126.1 [119.5]	54.4 [61.4]	1.1 [1.0]
Bahrain	198.8 [184.8]	216.6 [216.7]	8.6 [11.2]	0.6 [0.7]
Egypt	143.5 [102.7]	276.3 [261.3]	1.8 [1.9]	0.2 [0.2]
Iran	111.9 [76.1]	127.1 [118.0]	40.3 [46.8]	1.3 [1.2]
Iraq	163.3 [137.2]	162.8 [157.8]	26.6 [31.5]	1.2 [1.1]
Israel	80.7 [30.7]	138.8 [139.5]	54.1 [69.4]	1.3 [1.3]
Jordan	107.2 [60.6]	179.4 [172.5]	8.1 [12.8]	0.5 [0.6]
Kazakhstan	26.9 [10.3]	35.1 [34.4]	15.8 [13.5]	0.7 [0.7]
Kuwait	193.0 [171.7]	197.5 [194.3]	13.2 [21.5]	0.8 [1.0]
Kyrgyzstan	1.2 [0.1]	26.5 [24.4]	78.3 [76.7]	0.6 [0.6]
Lebanon	29.2 [7.5]	89.4 [91.9]	126.3 [158.6]	1.7 [1.6]
Oman	104.1 [60.3]	130.1 [142.7]	9.6 [6.8]	0.6 [0.5]
Pakistan	137.8 [105.9]	90.8 [90.5]	49.8 [47.3]	1.3 [1.3]
Qatar	199.2 [178.0]	232.8 [223.0]	12.7 [12.9]	0.9 [1.0]
Saudi Arabia	177.6 [137.8]	183.4 [180.7]	12.6 [11.3]	0.7 [0.6]
Syria	111.8 [77.2]	119.5 [118.0]	25.0 [29.9]	0.8 [0.8]
Tajikistan	17.2 [7.8]	49.2 [44.5]	133.2 [121.6]	0.8 [0.7]
Turkmenistan	98.1 [62.2]	102.8 [94.5]	8.8 [10.0]	0.7 [0.7]
UAE	202.6 [172.2]	226.3 [209.7]	10.6 [9.1]	0.8 [0.7]
Uzbekistan	75.5 [41.0]	82.7 [76.6]	13.5 [14.3]	0.8 [0.7]
Yemen	80.4 [38.3]	108.1 [103.4]	13.1 [9.9]	0.7 [0.6]

NOTE: Historical values are shown in brackets.

APPENDIX B

IABBREVIATIONS

AOR	area of responsibility
CENTCOM	U.S. Central Command
CFA	Cooperative Framework Agreement
CLM	Community Land Model
CMIP	Coupled Model Intercomparison Project
CO$_2$	carbon dioxide
DCAT	Department of Defense Climate Assessment Tool
DoD	U.S. Department of Defense
EEZ	exclusive economic zone
GCM	general circulation model
GDP	gross domestic product
GERD	Grand Ethiopian Renaissance Dam
IPCC	Intergovernmental Panel on Climate Change
ISIMIP	Inter-Sectoral Impact Model Intercomparison Project
NASA	National Aeronautics and Space Administration
OSMOSE	Object-oriented Simulator of Marine ecOSystEms
PET	potential evapotranspiration
RCP	Representative Concentration Pathway
UAE	United Arab Emirates

REFERENCES

Abbas, Nahlah, Sultana Nasrin, Nadhir Al-Ansari, and Sabah H. Ali, "The Impacts of Sea Level Rise on Basrah City, Iraq," *Open Journal of Geology*, Vol. 10, No. 12, December 2020.

Achakulwisut, Pattanun, Susan C. Anenberg, James E. Neumann, Stefani L. Penn, Natalie Weiss, Allison Crimmins, Neal Fann, Jeremy Martinich, Henry Roman, and Loretta J. Mickley, "Effects of Increasing Aridity on Ambient Dust and Public Health in the U.S. Southwest Under Climate Change," *GeoHealth*, Vol. 3, No. 5, May 2019.

Adeli, Afshin, "An Analysis of Employment and Fisheries Businesses Opportunities in Iran and the World," *Journal of Utilization and Cultivation of Aquatics*, Vol. 9, No. 1, April 2020.

Ahmed, Ehab Sh., and Ahmed S. Hassan, "The Impact of Extreme Air Temperatures on Characteristics of Iraq Weather," *Iraqi Journal of Science*, Vol. 59, No. 2C, March 2018.

Aldaya, M. M., G. Muñoz, and A. Y. Hoekstra, *Water Footprint of Cotton, Wheat and Rice Production in Central Asia*, UNESCO-IHE Institute for Water Education, March 2010.

Altz-Stamm, Amelia, *Jordan's Water Resource Challenges and the Prospects for Sustainability*, term paper for GIS for Water Resources, University of Texas at Austin, Fall 2012.

Asian and Pacific Centre for the Development of Disaster Information Management, *Sand and Dust Storms Risk Assessment in Asia and the Pacific*, August 2021.

Aw-Hassan, Aden, Fadel Rida, Roberto Telleria, and Adriana Bruggeman, "The Impact of Food and Agricultural Policies on Groundwater Use in Syria," *Journal of Hydrology*, Vol. 513, May 2014.

Beguería, Santiago, Borja Latorre, Fergus Reig, and Sergio M. Vicente-Serrano, "About the SPEI," webpage, undated. As of June 19, 2023: https://spei.csic.es/home.html

Bromberg, Gidon, Munqeth Mehyar, and Nader Khateeb, "The Jordan River," Middle East Institute, June 18, 2008.

Central Asian Bureau for Analytical Reporting, "Abnormal Heat in Central Asia: Society Must Take This Problem Seriously," 2021.

Chandler, Nathan, Jeffrey Martini, Karen M. Sudkamp, Maggie Habib, Benjamin J. Sacks, and Zohan Hasan Tariq, *Pathways from Climate Change to Conflict in U.S. Central Command*, RAND Corporation, RR-A2338-2, 2023.

Chaudhry, Iffat S., Asma H. Al-Harthi, Ghada M. Al-Shihimi, Khalsa M. Al-Saiti, and Maryam S. Al-Siyabi, "Millennials Outlook Towards Opportunities and Challenges in Fishing Industry of Muscat Governorate of Oman," *Middle East Journal of Management*, Vol. 4, No. 1, 2017.

Climate Central, "Coastal Risk Screening Tool," interactive map, undated. As of June 14, 2023: https://coastal.climatecentral.org/

"Climate Change Made Heatwaves in India and Pakistan '30 Times More Likely,'" World Meteorological Organization, May 24, 2022.

Columbia University, "Land Use and Agricultural Map of Iraq," webpage, undated. As of June 14, 2023: https://ciaotest.cc.columbia.edu/special_section/iraq_review/pi_map/pi_map_02.html

Crisp, Wil, "After Comeback, Southern Iraq's Marshes Are Now Drying Up," Yale Environment 360, January 10, 2023.

Department of the Army, Office of the Assistant Secretary of the Army for Installations, Energy and Environment, *United States Army Climate Strategy*, February 2022.

Directorate of Population and Manpower Statistics, *Iraq Population Estimates (2020)*, Central Bureau of Statistics Iraq, November 2020.

"Egypt Warns Ethiopia over Nile Dam," Al Jazeera, June 11, 2013.

El-Shirbeny, Mohammed A., and Khaled A. Abutaleb, "Monitoring of Water-Level Fluctuation of Lake Nasser Using Altimetry Satellite Data," *Earth Systems and Environment*, Vol. 2, No. 4, May 2018.

Fahim, Kareem, and Miriam Berger, "Protests over Water Shortages in Iran Turn Deadly in a Summer of Drought and Rolling Blackouts," *Washington Post*, July 21, 2021.

"Fisheries Sector Sees Growth of 44.5% in 2020," *Oman Daily Observer*, November 20, 2021.

Food and Agriculture Organization, *Irrigation Water Management: Irrigation Water Needs*, Training Manual No. 3, 1986.

Friedler, Eran, "Water Reuse—An Integral Part of Water Resources Management: Israel as a Case Study," *Water Policy*, Vol. 3, No. 1, 2001.

Gebremichael, Esayas, Mohamed Sultan, Richard Becker, Mohamed El Bastawesy, Omar Cherif, and Mustafa Emil, "Assessing Land Deformation and Sea Encroachment in the Nile Delta: A Radar Interferometric and Inundation Modeling Approach," *Journal of Geophysical Research: Solid Earth*, Vol. 123, No. 4, April 2018.

Hassan, Zahraa M., Monim H. Al-Jiboori, and Hazima M. Al-Abassi, "The Effect of the Extreme Heat Waves on Mortality Rates in Baghdad During the Period (2004–2018)," *Al-Mustansiriyah Journal of Science*, Vol. 31, No. 2, 2020.

Hassen, Tarek Ben, and Hamid El Bilali, "Chapter 26—Water Management in the Gulf Cooperation Council: Challenges and Prospects," *Current Directions in Water Scarcity Research*, Vol. 5, 2022.

Hirvela, Arto, *Effects of Climate Change to Balance of Power in the AOR*, Near East South Asia Center for Strategic Studies, July 13, 2021.

Intergovernmental Panel on Climate Change, "IPCC WGI Interactive Atlas," webpage, undated. As of June 13, 2023: http://interactive-atlas.ipcc.ch/

International Labour Organization, *Working on a Warmer Planet: The Impact of Heat Stress on Labour Productivity and Decent Work*, 2019.

International Labour Organization, *One Is Too Many: The Collection and Analysis of Data on Occupational Injuries in Qatar*, November 2021.

International Trade Administration, *Jordan—Country Commercial Guide*, December 14, 2022.

Inter-Sectoral Impact Model Intercomparison Project, "ISIMIP2b," database, last published January 31, 2021. As of June 16, 2023: https://www.isimip.org/protocol/2b/

IPCC—*See* Intergovernmental Panel on Climate Change.

"Iran Targets Doubling Aquaculture Production by 2025," *Tehran Times*, November 16, 2022.

ISIMIP—*See* Inter-Sectoral Impact Model Intercomparison Project.

Issanova, Gulnura, Jilili Abuduwaili, Azamat Kaldybayev, Oleg Semenov, and Tatiana Dedova, "Dust Storms in Kazakhstan: Frequency and Division," *Journal of the Geological Society of India*, Vol. 85, March 2015.

Iturbide, Maialen, Jesús Fernández, José Manuel Gutiérrez, Joaquín Bedia, Ezequiel Cimadevilla, Javier Díez-Sierra, Rodrigo Manzanas, Ana Casanueva, Jorge Baño-Medina, Josipa Milovac, Sixto Herrera, Antonio S. Cofiño, Daniel San Martín, Markel García-Díez, Mathias Hauser, David Huard, and Özge Yelekci, "Repository Supporting the Implementation of FAIR Principles in the IPCC-WGI Atlas," Zenodo, database, version v2.0-final, August 9, 2021. As of June 14, 2023: https://zenodo.org/records/5171760

Jägermeyr, Jonas, Christoph Müller, Alex C. Ruane, Joshua Elliott, Juraj Balkovic, Oscar Castillo, Babacar Faye, Ian Foster, Christian Folberth, James A. Franke, Kathrin Fuchs, Jose R. Guarin, Jens Heinke, Gerrit Hoogenboom, Toshichika Iizumi, Atul K. Jain, David Kelly, Nikolay Khabarov, Stefan Lange, Tzu-Shun Lin, Wenfeng Liu, Oleksandr Mialyk, Sara Minoli, Elisabeth J. Moyer, Masashi Okada, Meridel Phillips, Cheryl Porter, Sam S. Rabin, Clemens Scheer, Julia M. Schneider, Joep F. Schyns, Rastislav Skalsky, Andrew Smerald, Tommaso Stella, Haynes Stephens, Heidi Webber, Florian Zabel, and Cynthia Rosenzweig, "Climate Impacts on Global Agriculture Emerge Earlier in New Generation of Climate and Crop Models," *Nature Food*, Vol. 2, No. 11, November 2021.

Jiang, Jie, and Tianjun Zhou, "Agricultural Drought over Water-Scarce Central Asia Aggravated by Internal Climate Variability," *Nature Geoscience*, Vol. 16, No. 2, 2023.

Keynoush, Banafsheh, "With the Hope Line, Iran Aims to Boost Seawater Transfer to Fight Growing Drought," Middle East Institute, June 9, 2021.

Kimenyi, Mwangi S., and John Mukum Mbaku, "The Limits of the New 'Nile Agreement,'" Brookings Institution, April 28, 2015.

Kottek, M., J. Grieser, C. Beck, B. Rudolf, and F. Rubel, "World Map of the Köppen-Geiger Climate Classification Updated," webpage, World Maps of Köppen-Geiger Climate Classification, version June 2006. As of June 14, 2023: https://koeppen-geiger.vu-wien.ac.at/present.htm

Kulp, Scott A., and Benjamin H. Strauss, "New Elevation Data Triple Estimates of Global Vulnerability to Sea-Level Rise and Coastal Flooding," *Nature Communications*, Vol. 10, No. 1, October 2019.

Lachkar, Zouhair, Michael Mehari, Muchamad Al Azhar, Marina Lévy, and Shafer Smith, "Fast Local Warming Is the Main Driver of Recent Deoxygenation in the Northern Arabian Sea," *Biogeosciences*, Vol. 18, No. 20, October 2021.

Le Quesne, W. J. F., L. Fernand, T. S. Ali, O. Andres, M. Antonpoulou, J. A. Burt, W. W. Dougherty, P. J. Edson, J. El Kharraz, J. Glavan, R. J. Mamiit, K. D. Reid, A. Sajwani, and D. Sheahan, "Is the Development of Desalination Compatible with Sustainable Development of the Arabian Gulf?" *Marine Pollution Bulletin*, Vol. 173, Part A, December 2021.

Liu, Yang, Xiu Geng, Zhixin Hao, and Jingyun Zheng, "Changes in Climate Extremes in Central Asia Under 1.5 and 2 °C Global Warming and Their Impacts on Agricultural Productions," *Atmosphere*, Vol. 11, No. 10, 2020.

Loveluck, Louisa, and Mustafa Salim, "Iraq Broils in Dangerous 120-Degree Heat as Power Grid Shuts Down," *Washington Post*, August 7, 2022.

Madani, Kaveh, "Water Management in Iran: What Is Causing the Looming Crisis?" *Journal of Environmental Studies and Sciences*, Vol. 4, No. 4, December 2014.

Masson-Delmotte, Valérie, Panmao Zhai, Anna Pirani, Sarah L. Connors, Clotilde Péan, Yang Chen, Leah Goldfarb, Melissa I. Gomis, J. B. Robin Matthews, Sophie Berger, Mengtian Huang, Ozge Yelekçi, Rong Yu, Baiquan Zhou, Elisabeth Lonnoy, Thomas K. Maycock, Tim Waterfield, Katherine Leitzell, and Nada Caud, eds., *Climate Change 2021: The Physical Science Basis. Contribution of Working Group I to the Sixth Assessment Report of the Intergovernmental Panel on Climate Change*, Intergovernmental Panel on Climate Change, Cambridge University Press, 2021.

McDonnell, Tim, "The Nile Delta Isn't Ready for Climate Change," *Quartz*, October 27, 2022.

McSweeney, Carol F., and Richard G. Jones, "How Representative Is the Spread of Climate Projections from the 5 CMIP5 GCMs Used in ISI-MIP?" *Climate Services*, Vol. 1, March 2016.

Mera, Getachew Alem, "Drought and Its Impacts in Ethiopia," *Weather and Climate Extremes*, Vol. 22, October 2018.

Meshkati, Najmedin, "Gulf Escalation Threatens Drinking Water," Belfer Center for Science and International Affairs, Harvard Kennedy School, June 26, 2019.

Müller Schmied, Hannes, Denise Cáceres, Stephanie Eisner, Martina Flörke, Claudia Herbert, Christoph Niemann, Thedini Asali Peiris, Eklavyya Popat, Felix Theodor Portmann, Robert Reinecke, Maike Schumacher, Somayeh Shadkam, Camelia-Eliza Telteu, Tim Trautmann, and Petra Döll, "The Global Water Resources and Use Model WaterGAP v2.2d: Model Description and Evaluation," *Geoscientific Model Development*, Vol. 14, No. 2, February 2021.

NASA—*See* National Aeronautics and Space Administration.

National Aeronautics and Space Administration, "Sea Level Projection Tool," webpage, undated. As of June 14, 2023: https://sealevel.nasa.gov/ipcc-ar6-sea-level-projection-tool

National Research Council, *Water Conservation, Reuse, and Recycling: Proceedings of an Iranian-American Workshop*, National Academies Press, 2005.

National Security, Military, and Intelligence Panel on Climate Change, *A Security Threat Assessment of Global Climate Change: How Likely Warming Scenarios Indicate a Catastrophic Security Future*, Center for Climate and Security, February 2020.

National Weather Service, "What Is the Heat Index?" webpage, undated. As of June 14, 2023: https://www.weather.gov/ama/heatindex

Nikiel, Catherine A., and Elfatih A. B. Eltahir, "Past and Future Trends of Egypt's Water Consumption and Its Sources," *Nature Communications*, Vol. 12, No. 1, July 2021.

Oleson, Keith, David M. Lawrence, Gordon B. Bonan, B. A. Drewniak, Maoyi Huang, Charles D. Koven, Samuel Levis, Fang Li, William J. Riley, Zachary M. Subin, Sean Swenson, Peter E. Thornton, Anil Bozbiyik, Rosie Fisher, Colette L. Heald, Erik Kluzek, Jean-Francois Lamarque, Peter J. Lawrence, L. Ruby Leung, William Lipscomb, Stefan P. Muszala, Daniel M. Ricciuto, William J. Sacks, Ying Sun, Jinyun Tang, and Zong-Liang Yang, *Technical Description of Version 4.5 of the Community Land Model (CLM)*, NCAR Technical Note No. NCAR/TN-503+ STR, National Center for Atmospheric Research, July 2013.

Paparella, Francesco, Daniele D'Agostino, and John A. Burt, "Long-Term, Basin-Scale Salinity Impacts from Desalination in the Arabian/Persian Gulf," *Scientific Reports*, Vol. 12, November 2022.

Pinson, A. O., K. D. White, E. E. Ritchie, H. M. Conners, and J. R. Arnold, *DoD Installation Exposure to Climate Change at Home and Abroad*, U.S. Army Corps of Engineers, April 2021.

Pörtner, Hans-Otto, Debra C. Roberts, Melinda M. B. Tignor, Elvira Poloczanska, Katja Mintenbeck, Andrés Alegría, Marlies Craig, Stefanie Langsdorf, Sina Löschke, Vincent Möller, Andrew Okem, and Bardhyl Rama, eds., *Climate Change 2022: Impacts, Adaptation, and Vulnerability. Contribution of Working Group II to the Sixth Assessment Report of the Intergovernmental Panel on Climate Change*, Intergovernmental Panel on Climate Change, Cambridge University Press, 2022.

Pourkerman, Majid, Nick Marriner, Sedigheh Amjadi, Razyeh Lak, Mohammadali Hamzeh, Gholamreza Mohammadpor, Hamid Lahijani, Morteza Tavakoli, Christophe Morhange, and Majid Shah-Hosseini, "The Impacts of Persian Gulf Water and Ocean-Atmosphere Interactions on Tropical Cyclone Intensification in the Arabian Sea," *Marine Pollution Bulletin*, Vol. 188, March 2023.

Qin, Yue, John T. Abatzoglou, Stefan Siebert, Laurie S. Huning, Amir AghaKouchak, Justin S. Mankin, Chaopeng Hong, Dan Tong, Steven J. Davis, and Nathaniel D. Mueller, "Agricultural Risks from Changing Snowmelt," *Nature Climate Change*, Vol. 10, No. 5, April 2020.

Regional Initiative for the Assessment of Climate Change Impacts on Water Resources and Socio-Economic Vulnerability in the Arab Region, *Arab Climate Change Assessment Report*, United Nations Economic and Social Commission for Western Asia, 2017.

Samir, Salwa, "Extreme Heat Takes Toll on Egypt's Archaeological Heritage," *Al-Monitor*, September 2, 2021.

Scheffran, Jürgen., P. Michael Link, and Janpeter Schilling, "Theories and Models of the Climate Security Link," Working Paper CLISEC-3, University of Hamburg, Research Group Climate Change and Security, January 2009.

Shaban, Amin, "Striking Challenges on Water Resources of Lebanon," in Muhammad Salik Javaid, ed., *Hydrology: The Science of Water*, InTechOpen, 2019.

Shahin, Mamdouh, "The Nubian Sandstone Basin in North Africa, A Source of Irrigation Water for Desert Oases," *Subsurface-Water Hydrology: Proceedings of the International Conference on Hydrology and Water Resources, New Delhi, India, December 1993*, Springer, 1996.

Shatz, Howard J., Karen M. Sudkamp, Jeffrey Martini, Mohammad Ahmadi, Derek Grossman, and Kotryna Jukneviciute, *Mischief,*

Malevolence, or Indifference? How Competitors and Adversaries Could Exploit Climate-Related Conflict in the U.S. Central Command Area of Responsibility, RAND Corporation, RR-A2338-4, 2023.

Shin, Yunne-Jai, and Philippe Cury, "Exploring Fish Community Dynamics Through Size-Dependent Trophic Interactions Using a Spatialized Individual-Based Model," *Aquatic Living Resources*, Vol. 14, No. 2, March 2001.

Spiritos, Erica, and Clive Lipchin, "Desalination in Israel," in Nir Becker, ed., *Water Policy in Israel: Context, Issues, and Options*, Springer, 2013.

Sudkamp, Karen M., Elisa Yoshiara, Jeffrey Martini, Mohammad Ahmadi, Matthew Kubasak, Alexander Noyes, Alexandra Stark, Zohan Hasan Tariq, Ryan Haberman, and Erik E. Mueller, *Defense Planning Implications of Climate Change for U.S. Central Command*, RAND Corporation, RR-A2338-5, 2023.

Swedish Meteorological and Hydrological Institute and United Nations Economic and Social Commission for Western Asia, *Future Climate Projections for the Mashreq Region: Summary Outcomes*, Regional Initiative for the Assessment of the Impact of Climate Change on Water Resources and Socio-Economic Vulnerability in the Arab Region, 2021.

Tian, Jing, and Yongqiang Zhang, "Detecting Changes in Irrigation Water Requirement in Central Asia Under CO2 Fertilization and Land Use Changes," *Journal of Hydrology*, Vol. 583, April 2020.

Toukan, Mark, Stephen Watts, Emily Allendorf, Jeffrey Martini, Karen M. Sudkamp, Nathan Chandler, and Maggie Habib, *Conflict Projections in U.S. Central Command: Incorporating Climate Change*, RAND Corporation, RR-A2338-3, 2023.

UNICEF Jordan, "Water, Sanitation, and Hygiene," webpage, undated. As of June 14, 2023:
https://www.unicef.org/jordan/water-sanitation-and-hygiene

United Nations Economic and Social Commission for Western Asia and Bundesanstalt für Geowissenschaften und Rohstoffe, *Inventory of Shared Water Resources in Western Asia*, 2013.

United Nations Iraq, "Iraq on Track in the Preparation of Its Climate Change National Adaptation Plan," press release, July 28, 2022.

U.S. Agency for International Development, *The Fisheries Sector in Yemen: Status and Opportunities*, November 2019.

van Vuuren, Detlef P., Jae Edmonds, Mikiko Kainuma, Keywan Riahi, Allison Thomson, Kathy Hibbard, George C. Hurtt, Tom Kram, Volker Krey, Jean-Francois Lamarque, Toshihiko Masui, Malte Meinshausen, Nebojsa Nakicenovic, Steven J. Smith, and Steven K. Rose, "The Representative Concentration Pathways: An Overview," *Climatic Change*, Vol. 109, No. 1, August 2011.

Wahab, Engy Abdel, "How 'Climate-Smart' Crops Could Prove a Lifeline for Vulnerable Smallholders on the Nile Delta," United Nations Development Programme, May 23, 2022.

World Bank, "Arable Land (% of Land Area)," database, undated. As of June 13, 2023:
https://data.worldbank.org/indicator/AG.LND.ARBL.ZS?name_desc=false

World Bank, "Labor Force, Total—Iran, Islamic Rep.," database, 2022.

World Bank Group, *Egypt Country Climate and Development Report*, CCDR Series, 2022a.

World Bank Group, *Iraq Country Climate and Development Report*, CCDR Series, 2022b.

World Bank Group, *Jordan Country Climate and Development Report*, CCDR Series, 2022c.

World Bank Group, *Kazakhstan Country Climate and Development Report*, CCDR Series, 2022d.

World Bank Group, *Pakistan Country Climate and Development Report*, CCDR Series, 2022e.

Yoon, Jim, Christian Klassert, Philip Selby, Thibaut Lachaut, Stephen Knox, Nicolas Avisse, Julien Harou, Amaury Tilmant, Bernd Klauer, Daanish Mustafa, Katja Sigel, Samer Talozi, Erik Gawel, Josue Medellín-Azuara, Bushra Bataineh, Hua Zhang, and Steven M. Gorelick, "A Coupled Human–Natural System Analysis of Freshwater Security Under Climate and Population Change," *Proceedings of the National Academy of Sciences*, Vol. 118, No. 14, April 2021.

Yuan, Shawn, "Heatwaves Scorch Iraq as Protracted Political Crisis Grinds On," Al Jazeera, August 6, 2022.

Zaidan, Esmat, Mohammad Al-Saidi, and Suzanne H. Hammad, "Sustainable Development in the Arab World—Is the Gulf Cooperation Council (GCC) Region Fit for the Challenge?" *Development in Practice*, Vol. 29, No. 5, July 2019.

Zhu, Zaichun, Shilong Piao, Ranga B. Myneni, Mengtian Huang, Zhenzhong Zeng, Josep G. Canadell, Philippe Ciais, Stephen Sitch, Pierre Friedlingstein, Almut Arneth, Chunxiang Cao, Lei Cheng, Etsushi Kato, Charles Koven, Yue Li, Xu Lian, Yongwen Liu, Ronggao Liu, Jiafu Mao, Yaozhong Pan, Shushi Peng, Josep Peñuelas, Benjamin Poulter, Thomas A. M. Pugh, Benjamin D. Stocker, Nicolas Viovy, Xuhui Wang, Yingping Wang, Zhiqiang Xiao, Hui Yang, Sönke Zaehle, and Ning Zeng, "Greening of the Earth and Its Drivers," *Nature Climate Change*, Vol. 6, No. 8, April 2016.

REFERENCES